Fundamentals of
Radiation Dosimetry

Medical Physics Handbooks
Other books in the series

Series Editor: **Professor J M A Lenihan**
Department of Nursing Studies
University of Glasgow

Medical Physics Handbooks 15

Fundamentals of Radiation Dosimetry

J R Greening
Department of Medical Physics and
Medical Engineering, University of
Edinburgh and South East Scotland
Health Boards

Second Edition

CRC Press
Taylor & Francis Group
Boca Raton London New York

CRC Press is an imprint of the
Taylor & Francis Group, an **informa** business
A TAYLOR & FRANCIS BOOK

First published 1985 Taylor & Francis Group

Published 2019 by CRC Press
Taylor & Francis Group
6000 Broken Sound Parkway NW, Suite 300
Boca Raton, FL 33487-2742

© 1985 by J R Greening
CRC Press is an imprint of Taylor & Francis Group, an Informa business

First issued in paperback 2019

No claim to original U.S. Government works

ISBN-13: 978-0-367-45173-8 (pbk)
ISBN-13: 978-0-8527-4789-6 (hbk)

**Visit the Taylor & Francis Web site at
http://www.taylorandfrancis.com**

**and the CRC Press Web site at
http://www.crcpress.com**

Library of Congress Cataloging-in-Publication Data

Catalog record is available from the Library of Congress

To my Wife and Sons

Contents

Preface to the Second Edition

Since the preparation of the first edition protocols for the dosimetry of photons have been issued by the Nordic Association of Clinical Physicists (NACP), the Hospital Physicists' Association and the American Association of Physicists in Medicine (AAPM). Other protocols for electrons have been published by NACP, AAPM and the International Commission on Radiation Units and Measurements (ICRU), and protocols for neutrons by both AAPM and European dosimetrists. Also the ICRU is about to publish (early 1985) recommendations of new quantities for use in radiation protection that should help substantially to improve statements of dose equivalent. These important publications have led to appreciable changes in Chapters 9 and 12.

The opportunity has been taken of incorporating a number of helpful suggestions from international correspondents and of making other improvements and minor corrections. Appendices have been added to summarise some details of radiation quantities and the relationships between them. However, I have resisted the temptation to expand the book greatly and run the risk of confusing the reader with detail before the fundamentals have been understood. That detail can be pursued as needed in the references, which have been updated throughout and now number almost 350.

J R Greening

Preface to the First Edition

Newcomers to the field of radiation dosimetry are confronted by several quantities and associated units which they may well not have encountered in other fields of science. This book defines those quantities and the relationships between them and discusses the principles underlying their measurement. The definitions follow those of the latest (1980) Quantities and Units Report of the International Commission on Radiation Units and Measurements (ICRU), an inexpensive publication strongly recommended to all serious students of dosimetry. The book is published at a time of transition from special radiation units to SI units in radiation dosimetry. Accordingly almost all equations have been expressed as relationships between quantities and are true whatever the units employed. However, specific examples are given of the effects of changing units.

The book has been written with physicists in mind, but the treatment is such that most of it should be accessible to those trained in other disciplines. It has been necessary to be selective both of topics and the depth of their discussion. There has been concentration on fundamentals, and it is hoped that a fairly generous use of references will lead the reader to deeper discussions of particular topics in the authoritative reports of the ICRU, chapters in the major three-volume work 'Radiation Dosimetry', and individual publications.

The symbolism and nomenclature used by the ICRU have been adopted as far as possible. Although these are largely a matter of convention it is helpful if there is common usage and the example of an international body seemed the one to follow. The labelling of graphs and tables follows the practice of the ICRU, the Royal Society and some Europhysics journals. If it is unfamiliar there is a brief explanation at the end of the Appendix.

The text has benefited greatly from the comments of Dr H O Wyckoff and several of my departmental colleagues. I am most grateful to them all. I also wish to record my special thanks to my secretary, Mrs D Y Nicholson, for her care in the preparation of the typescript and her tolerance of textual changes.

J R Greening

1 The Radiation Field

1.1 Introduction

Most quantitative measurements of ionising radiation quantities are undertaken either to establish or to use numerical relationships between them and the biological, physical or chemical effects the radiations produce. These effects can only occur with the transfer of energy from the radiation to some irradiated material. The effect is likely to be different if a particular amount of energy is imparted to a small mass of material rather than being distributed throughout a large mass. The most widely used dosimetric quantity is therefore this imparted energy divided by the mass concerned. It is called *absorbed dose* and is more precisely defined in Chapter 4. Most of this book is devoted to the determination of absorbed dose and related dosimetric quantities. However, the dosimetric quantities arise through the interaction of some property of a radiation field with a material. Indeed dosimetric quantities can be expressed as the product of a radiation field quantity and an interaction coefficient, so before considering dosimetric quantities attention must be given to radiation fields and interaction coefficients.

1.2 Radiation Source

A source of ionising radiation such as an x-ray tube, electron accelerator or radioactive material may be characterised by its rate of emission of ionising particles. These may be directly ionising as, for example, with charged particles such as electrons and α particles or indirectly ionising as with uncharged particles such as photons and neutrons. These latter particles release electrons or recoil nuclei and other charged particles, respectively, which then cause ionisation. The characterisation of the source can be refined by adding further information such as the distribution of these particles in energy, direction and time.

1.3   Radiation Field Quantities

A radiation source will give rise to a radiation field. Within this field there will be a *fluence*, Φ, of particles defined by the International Commission on Radiation Units and Measurements (ICRU 1980) as $\Phi = dN/da$ where dN is the number of particles incident on a sphere of cross sectional area da. Reference to a sphere with a certain cross sectional area rather than to the cross sectional area itself removes the need to specify the orientation of the cross sectional area and the definition applies equally well to mono-directional or multi-directional particles. The SI unit of fluence is m^{-2}. A further quantity, the *fluence rate*, ϕ, is defined as $d\Phi/dt$ where $d\Phi$ is the increment of fluence in the time interval dt. The SI unit is $m^{-2}s^{-1}$. It is somewhat unusual for a rate quantity to be given its own distinctive symbol, but many people regard fluence rate as being a more fundamental property of a radiation field than is the fluence. The same comment applies to *energy fluence rate* defined below.

Consideration may be given to the energy carried by the particles rather than to the particles themselves. The *energy fluence*, Ψ, is defined as $\Psi = dR/da$ where dR is the radiant energy entering a sphere of cross sectional area da. The SI unit of energy fluence is Jm^{-2}. The *radiant energy*, R, is the energy, excluding rest energy, of the particles concerned. The *energy fluence rate*, ψ, is $d\Psi/dt$ where $d\Psi$ is the increment of energy fluence in the time interval dt. The SI unit is Wm^{-2}.

Care should be taken not to confuse the fluence and energy fluence as defined above in terms of an elementary sphere with the corresponding quantities, sometimes called the *plane* or *planar fluence* and *plane* or *planar energy fluence*, defined in terms of an elementary area da of a plane of fixed orientation. In the latter cases the particle or energy transport is considered positive when incident on one side of the plane and negative when incident on the other. The plane or planar fluence and plane or planar energy fluence are zero for isotropic radiation (see Carlsson 1979).

Several other radiation field quantities have been defined (ICRU 1980) but those given above are all that are needed to derive the dosimetric quantities discussed later.

1.4   Distributions of Field Quantities

The specification of a radiation field is more complete if the distributions of fluence or energy fluence in direction and energy can be given. An adequate knowledge of the distribution in direction often follows from knowing the position of the initial source of radiation. The direction of secondary radiations arising from the interaction of primary radiation with matter may be determined from the detailed theory of the interaction process and some examples are given in figures 2.2, 2.4 and 2.5.

Interest is more likely to be focused on the distribution of the radiation in space resulting from its distribution in direction. In any event, should the distribution in direction be required it can be measured by using a collimated detector that only accepts radiation coming from a limited solid angle.

The distribution of fluence or energy fluence with respect to energy is a much more important matter. The response of a detector to a particular particle fluence is frequently a function of the energy distribution of that fluence, and in consequence a substantial effort may be put into determining this energy distribution either by measurement or calculation. It is usual to employ the differential distribution of the quantity with respect to energy, namely Φ_E where $\Phi_E \, dE$ is the fluence of particles with energies lying between E and $E + dE$. The total fluence, Φ, is given by

$$\Phi = \int_0^{E_{\max}} \Phi_E \, dE, \qquad (1.1)$$

and corresponds to the area under the curve of figure 1.1 which shows the differential distribution in energy of the fluence of β particles from the radionuclide ^{114}In.

1.5 Mean or 'Effective' Energy

It is possible to calculate the mean energy, $\bar{E}$, of the particles weighted by fluence from

$$\bar{E} = \frac{\displaystyle\int_0^{E_{\max}} E \, \Phi_E \, dE}{\displaystyle\int_0^{E_{\max}} \Phi_E \, dE}. \qquad (1.2)$$

If, however, each particle contributing to the total fluence is multiplied by its energy we get the differential distribution of *energy* fluence with respect to energy, Ψ_E.

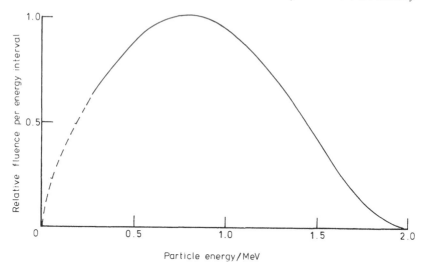

Figure 1.1   Differential distribution of β particle fluence with energy for ^{114}In.

A mean energy based on a weighting by energy fluence is obtained from

$$\bar{E} = \frac{\displaystyle\int_0^{E_{\max}} E\Psi_E\,\mathrm{d}E}{\displaystyle\int_0^{E_{\max}} \Psi_E\,\mathrm{d}E}. \qquad (1.3)$$

But in general $\Psi_E/\Psi \neq \Phi_E/\Phi$, so equations (1.2) and (1.3) lead to different values of $\bar{E}$. Thus in expressing the mean energy of any radiation it is necessary to state the quantity whose differential energy distribution has been used in calculating the mean. The problem can be more complicated than this. With photons, for example, it will be seen later that we can measure other quantities such as absorbed dose in various materials (see Chapter 4), or exposure (see Chapter 5). These quantities are obtained from energy fluence by multiplying by some of the energy dependent interaction coefficients of Chapter 2. These additional quantities can give rise to other values for the mean energy $\bar{E}$ when $\bar{E}$ is calculated from further equations corresponding to equations (1.2) and (1.3). Caution has to be exercised in attributing to a beam of radiation a mean or 'effective' energy. The nature of the mean or the respect in which the 'effectiveness' operates should be stated. Figure 1.2 illustrates the differential distribution in energy of fluence, energy fluence and exposure for the *same* photon beam, namely a typical diagnostic radiology beam generated at 70 kV with 2 mm Al filter. It will be seen that the spectral distributions of the three quantities are substantially different and would give rise to different mean values for the photon energies. Furthermore, the thickness of an absorber that reduces the photon beam to half its initial value—called the half-value thickness (HVT) or half-value layer (HVL)—will also depend on the quantity that the detector measures.

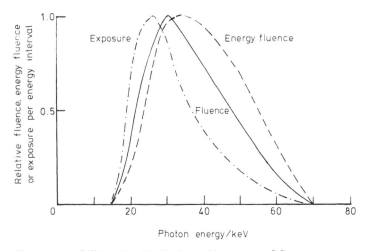

Figure 1.2 Differential distribution with energy of fluence, energy fluence and exposure for x-ray beam generated at 70 kV with 2 mm Al filter.

Another example is given in figure 3.6 which shows the considerable difference between the differential distributions in energy of ^{252}Cf neutrons when expressed in terms of fluence and energy fluence.

Problems of radiation measurement are considerably eased, and the need for a detailed knowledge of the spectral distribution of the radiation is removed, if the response of the radiation detector to a particular quantity does not vary, or varies only slightly, within the range of energies covered by the measurement. This point is discussed further in Chapter 3 where, in addition, the determination of spectral distributions is reviewed.

2 Interactions of Ionising Radiations with Matter

2.1 Cross Sections and Interaction Coefficients

In the context of indirectly ionising radiations, the word interaction is applied to processes in which the energy and/or the direction of the radiation is altered. Such processes are random and it is therefore only possible to speak of the probability of interactions occurring. This probability can be expressed in terms of cross sections or of various interaction coefficients.

Thus the concept can be introduced of a cross section or *apparent* area that an interaction centre (electron, nucleus, atom, etc) presents to the radiation which, if traversed by the radiation, gives rise to an interaction. More precisely in terms of probability

$$\text{cross section} = \frac{\text{probability of interaction}}{\text{unit particle fluence}}. \tag{2.1}$$

This cross section is usually represented by the symbol σ, and has dimensions $[L^2]$. The SI unit is therefore m^2 but a special unit, the barn, b, equal to 10^{-28} m^2 is also used.

Suppose B is the number of interaction centres per unit volume and the fluence Φ is mono-directional, then there will be $B\sigma\Phi$ interactions per unit path length or $B\sigma\Phi dl$ in a path length dl. The fractional change of fluence, $d\Phi/\Phi$, in path dl is then $-B\sigma dl$. $B\sigma$ is called the linear attenuation coefficient, is usually represented by the symbol μ, and has dimension $[L^{-1}]$. The SI unit is m^{-1}. Thus

$$\frac{d\Phi}{\Phi} = -\mu\,dl \qquad \text{and} \qquad \Phi_l = \Phi_0 \exp(-\mu l), \tag{2.2}$$

where Φ_0 is the initial fluence and Φ_l is the fluence after traversing a path length l.

6

The number of atoms per unit volume of a substance of density ρ and molar mass M is $N_A \rho/M$, where N_A is the Avogadro constant. Thus $\mu = N_A \rho \sigma/M$, where σ is the atomic cross section.

It is also possible to express interaction probabilities in terms of the number of interaction centres per unit mass. This is obviously related to the number of interaction centres per unit volume by the factor $1/\rho$. By considerations similar to those given above we obtain a mass attenuation coefficient μ/ρ, which is the probability of interaction per unit of path length expressed in terms of mass/area, and has dimensions $[L^2 M^{-1}]$. The SI unit is $m^2 kg^{-1}$.

The mass attenuation coefficient is related to the atomic cross section σ by $\mu/\rho = N_A \sigma/M$.

We will discuss some partial mass attenuation coefficients of importance to radiation dosimetry after considering the major photon interaction processes.

2.2 Interactions of Photons with Matter

Photons can, in principle, interact with planetary electrons or atomic nuclei or the electric fields round these electrons or nuclei. In so doing they may lose all, part, or none of their energy. Of these nine possible processes, two are of great importance in medical applications, a third becomes important at photon energies above a few MeV, and two others need to be taken into account in limited circumstances.

The following discussion does not go into the detail of the theories of the interaction processes but sets out the terminology and reports the principal findings of theory and experiment.

2.2.1 Photoelectric effect

In this process a photon gives up all its energy $h\nu$ to a planetary electron which is then ejected from the atom with a kinetic energy $\frac{1}{2}mv^2$ (assuming non-relativistic conditions) equal to the energy of the photon less the energy E_b required to remove the electron from the attraction of the nucleus:

$$h\nu = \tfrac{1}{2}mv^2 + E_b.$$

Momentum is conserved by the recoil of the residual atom.

2.2.1.1   Variation with energy of photon. For this process to occur with a particular electron the energy of the photon must not be less than the binding energy, E_b, of the electron in the atom. For photon energies $h\nu > E_b$ the probability of interaction decreases as $h\nu$ increases, i.e. the probability is greatest when the energy of the photon and the binding energy of the electron are equal. In this energy region the photoelectric cross section per atom, $_{atom}\tau$, varies approximately as $(h\nu)^{-3}$. Over 80% of the primary interactions are in the K shell when the K shell can interact. When $h\nu < E_K$, the K shell binding energy, no K shell electrons can interact. There is therefore a sudden drop in interaction called the K absorption edge, as the photon energy decreases through E_K. For K absorption edge energies see table 2.1.

Similar effects occur at the three L absorption edges. These features are illustrated in figure 2.1.

At sufficiently low photon energies photoelectric absorption predominates over all other absorption processes.

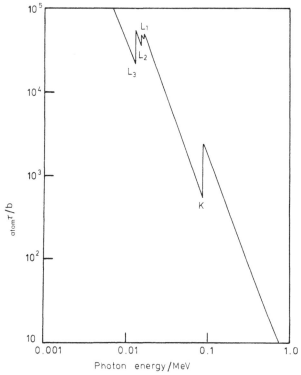

Figure 2.1 Photoelectric cross section per atom, $_{atom}\tau$, for lead as a function of photon energy.

Table 2.1 Fluorescence yields ω_K and K absorption edge energies (values of ω_K from Bambynek et al 1972).

Z	Element	K edge energy/keV	ω_K	Z	Element	K edge energy/keV	ω_K	Z	Element	K edge energy/keV	ω_K
3	Li	0.05		32	Ge	11.10	0.54	59	Pr	42.00	0.92
4	Be	0.11		33	As	11.87	0.57	60	Nd	43.57	0.92
6	C	0.28		34	Se	12.65	0.60	61	Pm	45.20	0.92
7	N	0.40		35	Br	13.47	0.62	62	Sm	46.85	0.93
8	O	0.53		36	Kr	14.32	0.65	63	Eu	48.52	0.93
10	Ne	0.87		37	Rb	15.20	0.67	64	Gd	50.23	0.93
11	Na	1.07	0.01	38	Sr	16.11	0.69	65	Tb	52.00	0.94
12	Mg	1.30	0.03	39	Y	17.04	0.71	66	Dy	53.79	0.94
13	Al	1.56	0.04	40	Zr	18.00	0.73	67	Ho	55.62	0.94
14	Si	1.84	0.05	41	Nb	18.99	0.75	68	Er	57.49	0.95
15	P	2.14	0.06	42	Mo	20.00	0.76	69	Tm	59.38	0.95
16	S	2.47	0.08	43	Tc	21.05	0.78	70	Yb	61.30	0.95
17	Cl	2.82	0.09	44	Ru	22.12	0.79	71	Lu	63.31	0.95
18	Ar	3.20	0.12	45	Rh	23.22	0.81	72	Hf	65.31	0.95
19	K	3.61	0.14	46	Pd	24.35	0.82	73	Ta	67.40	0.96
20	Ca	4.04	0.16	47	Ag	25.52	0.83	74	W	69.51	0.96
21	Sc	4.49	0.19	48	Cd	26.72	0.84	75	Re	71.66	0.96
22	Ti	4.96	0.22	49	In	27.94	0.85	76	Os	73.86	0.96
23	V	5.46	0.25	50	Sn	29.19	0.86	77	Ir	76.10	0.96
24	Cr	5.99	0.28	51	Sb	30.49	0.87	78	Pt	78.38	0.96
25	Mn	6.54	0.31	52	Te	31.81	0.88	79	Au	80.72	0.96
26	Fe	7.11	0.35	53	I	33.17	0.88	80	Hg	83.11	0.97
27	Co	7.71	0.38	54	Xe	34.59	0.89	81	Tl	85.53	0.97
28	Ni	8.33	0.41	55	Cs	35.99	0.90	82	Pb	88.01	0.97
29	Cu	8.98	0.44	56	Ba	37.45	0.90	83	Bi	90.53	0.97
30	Zn	9.66	0.48	57	La	38.93	0.91	90	Th	109.65	0.97
31	Ga	10.37	0.51	58	Ce	40.45	0.91	92	U	115.62	0.98

2.2.1.2   Variation with the atomic number of absorber. The photo-electric cross section per atom, $_{atom}\tau$, varies very roughly as Z^4. (But note that the photoelectric component τ/ρ of the mass attenuation coefficient μ/ρ varies approximately as Z^3, since $\tau/\rho = N_{A\ atom}\tau/M$ and M is approximately proportional to Z.) Thus high-Z materials are very strong absorbers of photons and are widely used for beam defining and radiation protection purposes, particularly with photons of low energy.

2.2.1.3   Direction of photoelectron emission. The direction of emission of photoelectrons varies with the energy of the incident photons. It will be seen from figure 2.2 that at very low energies the emission is predominantly at right angles to the direction of the incident photon, but as the photon energy rises the photoelectrons are increasingly emitted in the forward direction.

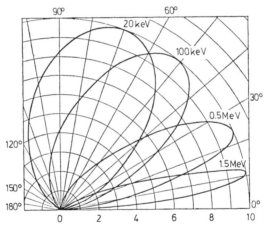

Figure 2.2   Angular distribution of photoelectrons (relative number per unit interval of angle) (from Whyte 1959).

2.2.1.4   Characteristic radiation. The ejection of a photoelectron from an atom leaves a vacancy in one of the electron orbits. This is filled by an electron falling in from an outer orbit. The potential energy given up by this electron sometimes appears as an x-ray photon. The photon energies are characteristic of the atom from which they come, and the photons are therefore called characteristic radiation. Photons resulting from electron transitions to the K shell are called K radiation, with corresponding names for the L and other shells. Characteristic radiation spectra have many lines, the K radiation comprises K_{α_1} (transition from L_{III} subshell to K shell), K_{α_2} (L_{II} to K), K_{β_1} which is a doublet (M_{II} and M_{III} to K) and K_{β_2} another doublet (N_{II} and N_{III} to K). These radiations can also arise when K shell vacancies are caused by charged particles, such as by electrons in an x-ray tube.

2.2.1.5 Auger effect. All electron transitions to vacancies in inner shells are not accompanied by characteristic radiation. Radiationless transitions occur in which the available energy is used to eject an electron from an outer shell. This process is called the Auger effect and the ejected electron is called an Auger electron. The atom now has two vacancies, which may be filled by the emission of further Auger electrons (an Auger cascade) leading to multiple ionisation of the atom. The possible transitions are often numerous and the spectra of Auger electrons can be very complex. (Radiationless transitions can also take place between subshells of shells such as the L and M. These are called Coster–Kronig transitions.)

2.2.1.6 Fluorescence yield. The fluorescence yield of an atomic shell or subshell is defined as the probability that a vacancy in that shell or subshell is filled through a radiative transition (i.e. with the emission of characteristic radiation). The probability of such transitions is approximately proportional to Z^4 while that of radiationless transitions is almost independent of Z. Thus the fluorescence yield of the K shell, ω_K, is given approximately by $Z^4/(a_K + Z^4)$. Burhop (1952) suggests $a_K = 1.12 \times 10^6$. Some values of ω_K recommended in an extensive review by Bambynek *et al* (1972) are given in table 2.1. It will be seen that low atomic number materials give rise to very little characteristic radiation. In any case it is of very low energy (about 0.5 keV for oxygen) and is reabsorbed locally.

2.2.2 Compton effect

In this process a photon of energy $h\nu_0$ (having wavelength λ_0 and momentum $h\nu_0/c$) collides with an electron which is regarded as free. (For this to be so the momentum transferred to the electron must greatly exceed its initial momentum within the atom.) The photon transfers some of its energy to the electron, which recoils, and the remainder of the energy $h\nu$ appears as the energy of a scattered photon of longer wavelength λ.

If we apply conservation of momentum and energy to the collision it may be shown that

$$\lambda - \lambda_0 = \frac{h}{mc} (1 - \cos \theta) = 0.024 (1 - \cos \theta) 10^{-10} \text{m} , \quad (2.3)$$

where θ is the angle between the directions of the incident and the scattered photons. Thus the *wavelength* change at a particular angle θ is independent of the initial photon wavelength (or energy). This is not the case for the energy change, which can be a very large fraction of the initial photon energy for high photon energies and is a very small fraction for low photon energies. The manner in which, *on average*, the incident photon energy is shared between the scattered photon and the recoil

Table 2.2 Average fraction of incident photon energy
retained by Compton scattered photon.

Incident photon energy/MeV	Average fraction retained by scattered photon
0.01	0.98
0.015	0.97
0.02	0.96
0.03	0.94
0.04	0.93
0.05	0.92
0.06	0.90
0.08	0.88
0.1	0.861
0.15	0.817
0.2	0.783
0.3	0.730
0.4	0.690
0.5	0.659
0.6	0.633
0.8	0.592
1.0	0.561
1.5	0.505
2	0.469
3	0.423
4	0.394
5	0.373
6	0.357
8	0.333
10	0.316

electron is shown in table 2.2. At any energy the particular sharing will depend on the scattering angle.

2.2.2.1 Variation with energy. This is illustrated in figure 2.3 where the total Compton cross section per electron, $_e\sigma$, is shown together with the partial cross sections $_e\sigma_s$ and $_e\sigma_a$, relating to the transfer of energy to scattered photons and recoil electrons, respectively. The total Compton cross section varies very little at the low energies used in x-ray diagnosis, but falls off at the higher energies used in modern radiotherapy.

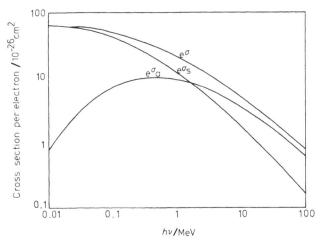

Figure 2.3 Variation of Compton cross sections with energy (from Whyte 1959).

2.2.2.2 Variation with atomic number. The cross section per electron, $_e\sigma$, is independent of Z. The cross section per atom, $_{atom}\sigma$, is given by multiplying the electronic cross section by the number of electrons per atom. Thus $_{atom}\sigma = {_e\sigma}Z$ and obviously is proportional to Z. The Compton component σ_c/ρ of the mass attenuation coefficient is proportional to $_{atom}\sigma/M$ (see §2.1) and therefore to Z/M. This varies only slowly with Z, except for the case of hydrogen for which it is about twice that of other low-Z elements.

2.2.2.3 Angular distribution of scattered photons. At very low energies the scattering is symmetrical about 90°, and as the photon energy increases the scattered energy tends more and more in the forward direction as illustrated in figure 2.4.

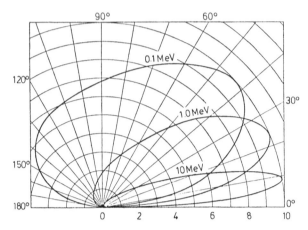

Figure 2.4 Angular distribution of Compton scattered photons (relative energy per unit interval of angle) (from Whyte 1959).

2.2.2.4 Angular distribution of Compton recoil electrons. Figure 2.5 shows that these electrons really are recoils, that is they are all ejected at angles of less than 90° from the direction of the initial photon. Otherwise, as with the scattered photons, they tend more and more in the forward direction as the photon energy increases.

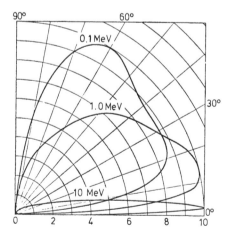

Figure 2.5 Angular distribution of Compton recoil electrons (relative number per unit interval of angle) (from Whyte 1959).

2.2.2.5   Energy distribution of Compton electrons. The maximum recoil electron energy, E_{max}, is given by

$$E_{max} = h\nu_0 \frac{2\,h\nu_0/mc^2}{1 + 2\,h\nu_0/mc^2}. \tag{2.4}$$

The average energy is $h\nu_0 \times {}_e\sigma_a/({}_e\sigma_a + {}_e\sigma_s)$.

The energy distribution for $h\nu_0 = 1$ MeV is of the form illustrated in figure 2.6. The sharp peak near E_{max} is detected by scintillation spectrometers and has to be taken into account in many clinical measurements of high-energy γ-ray emitters.

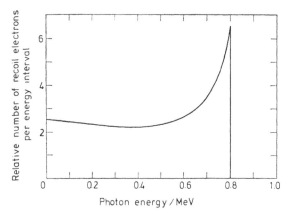

Figure 2.6   Energy distribution of Compton recoil electrons ($h\nu = 1$ MeV) (from Whyte 1959).

2.2.3   Pair production

In this process a photon interacts with the field round the nucleus of an atom, the whole of the photon's energy being converted to the mass and kinetic energy of a positive and a negative electron. The nucleus is necessary in order that momentum as well as energy may be conserved. No process is possible until the photon energy is equivalent to the mass of two electrons, i.e. to $2mc^2 = 1.02$ MeV.

2.2.3.1   Variation with energy. Once the threshold energy of 1.02 MeV is exceeded the probability of interaction increases as shown in figure 2.7.

2.2.3.2   Variation with atomic number. The pair production cross section per atom, ${}_{atom}\kappa$, varies approximately as Z^2. The degree of this approximation can be inferred from figure 2.7, where the ordinate is $_{atom}\kappa/Z^2$, by the extent to which the curves for the various elements fail to coincide. The pair production mass attenuation coefficient κ/ρ is approximately proportional to Z.

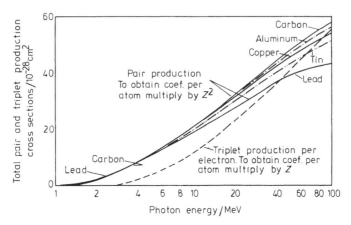

Figure 2.7   Pair and triplet cross sections as a function of photon energy for carbon, aluminium, copper, tin and lead. To obtain the coefficient per atom, multiply the triplet curve by Z and the pair production curves by Z^2 (from Johns and Laughlin 1956).

2.2.3.3   Positrons and electrons. The mean direction of the positron–electron pair is close to that of the photon. At high energies the average angle θ between the direction of the electron (or positron) and the incident photon is given by $\theta = mc^2/E$, where E is the energy of the electron.

All sharings of kinetic energy between the electron and positron are possible, those giving 20–80% to the electron (or positron) being about equally probable, with more uneven sharings becoming rapidly less likely.

2.2.3.4   Pair production in field of electron. This process is also possible, and is sometimes referred to as triplet production as the electron recoils giving a third energetic particle. The threshold for the reaction is $4mc^2 = 2.04$ MeV. The variation of triplet production with energy is shown in figure 2.7. The cross section per atom is obviously proportional to the number of electrons in the atom, i.e. to Z, and is substantially less than $_{atom}\kappa$, especially for high-Z materials.

2.2.3.5   Annihilation radiation. When the positron has slowed nearly to rest it annihilates with an electron giving two photons, each of energy $mc^2 = 0.51$ MeV, which travel in opposite directions. (Some in-flight annihilation also occurs giving photons with energy greater than 0.51 MeV.) Thus, although the incident-photon energy is $h\nu$, an energy of only $h\nu - 2mc^2$ is transferred to kinetic energy of charged particles, and an energy of $2mc^2$ may well escape from an irradiated medium.

2.2.4   Rayleigh (coherent) scattering

Consideration will now be given to two interaction processes of less importance than the three discussed so far, but which need to be allowed for in some circumstances.

The first of these is Rayleigh scattering. In this process the incident photon collides with an electron which is sufficiently tightly bound in an atom for the whole atom to absorb the recoil. The energy transfer to the atom is then negligible and the photon is scattered without loss of energy. Furthermore, the scattering contributions from the electrons in a given atom will all have a definite phase relationship to each other. In the forward direction the contribution will interfere constructively so that the amplitude will be Z times and the energy fluence rate Z^2 times that due to a single electron. As the scattering angle increases the interference becomes destructive and the scattered energy fluence rate falls off very substantially. The scattered radiation is very strongly peaked in the forward direction.

2.2.4.1   Variation with energy and atomic number. The Rayleigh-scattering cross section per atom, $_{\text{atom}}\sigma_{\text{coh}}$, varies approximately as follows

$$_{\text{atom}}\sigma_{\text{coh}} \propto (h\nu)^{-2} Z^{2.5}. \tag{2.5}$$

Generalisations about the relative importance of Rayleigh scattering are difficult, but figure 2.8 illustrates the percentage of the total attenuation cross section which is due to this process. In the case of carbon, which may be taken as typical of the low atomic number materials of importance in biology and radiation dosimetry, Rayleigh scattering contributes a maximum of 15% to the total attenuation cross section, and this occurs at about 20 keV.

2.2.5   Nuclear photoeffect

The second minor process to be considered is that in which a photon interacts with the nucleus of an atom, which subsequently emits energy

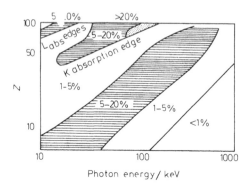

Figure 2.8 Percentage of Rayleigh scattering in the total attenuation cross section, as a function of Z and photon energy (from Dyson 1973).

usually in the form of neutrons or protons. The effect, which has the form of a giant resonance, sets in at thresholds of a few MeV, rises to a maximum at 12–24 MeV and falls off rapidly at higher energies. It rarely amounts to more than 5% of the total attenuation cross section and then only in a limited energy range around 15 to 20 MeV.

Tabulations of cross sections tend to omit the nuclear photoeffect and for this reason some indication of its importance relative to the three main interaction processes is given in table 2.3. Attention should be given to this effect if precise data are being prepared for accurate dosimetry at photon energies around 20 MeV.

2.2.6 Relative importance of photoelectric, Compton and pair production processes

Photoelectric interactions predominate for all materials at sufficiently low photon energies, but as the energy increases the photoelectric effect falls off more rapidly than the Compton effect and the latter eventually takes over as the dominating process. With continuing increase of photon energy, although the Compton effect decreases absolutely, it increases still more relative to the photoeffect, but when photon energies reach several MeV the pair production process begins to make the greatest contribution to photon interactions. The situation is illustrated in figure 2.9 for (*a*) a material, carbon, typical of the atomic numbers of importance in biology and dosimetry, and for (*b*) lead, which is a representative high atomic number material. It should be noted that for the typical low atomic number material the Compton process predominates from about

Table 2.3 Nuclear photoeffect cross sections, $\sigma_{\gamma,\text{nucl}}$ (from Hubbell 1969, based on data provided by E G Fuller).

Nucleus	Threshold energy/MeV (γ, n)	(γ, p)	Energy/MeV at $\sigma_{\gamma,\text{nucl}}$ peak	Maximum $_{\text{atom}}\sigma_{\gamma,\text{nucl}}$/b	Peak width at half maximum/MeV	Maximum $\sigma_{\gamma,\text{nucl}}$ as percentage of other cross sections
^{12}C	18.7	16.0	23	0.018	3.6	5.9
^{27}Al	13.1	8.3	21.5	0.038	9.0	3.9
^{40}Ca	15.7	8.3	20.5	0.100	4.5	5.2
^{63}Cu	10.8	6.1	17.0	0.070	8.0	2.0
^{90}Zr	12.0	8.4	17.0	0.180	4.5	3.0
^{127}I	9.1	6.2	15.2	0.210	5.7	2.3
^{165}Ho	8.0	6.1	14.0	0.220	8.5	1.7
^{181}Ta	7.6	6.2	14.0	0.280	6.5	1.8
^{108}Pb	7.4	8.0	13.6	0.495	3.8	2.7
^{235}U	6.1	7.6	12.2	0.500	7.0	2.4

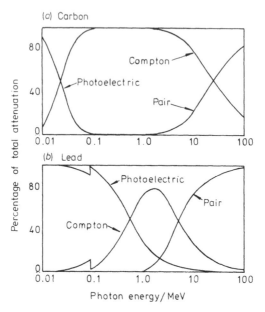

Figure 2.9 Relative probability of different effects for photons of different energies in carbon and lead.

25 keV to 25 MeV, an energy range covering almost all current applications of x- and γ rays in medicine.

The small relative contributions of Rayleigh scattering and the photonuclear effect have been discussed earlier.

2.3 Interaction Coefficients

2.3.1 Total interaction coefficients

The probabilities of interactions by the processes so far discussed are independent of one another and in consequence the total interaction probability is the sum of the individual interaction probabilities. For example, the total mass attenuation coefficient is the sum of the mass attenuation coefficients for the individual interactions

$$\frac{\mu}{\rho} = \frac{\tau}{\rho} + \frac{\sigma_c}{\rho} + \frac{\kappa}{\rho} + \frac{\sigma_{coh}}{\rho} + \frac{\tau_{nucl}}{\rho}. \tag{2.6}$$

2.3.1.1 Interaction coefficients for mixtures and compounds. The interaction coefficient of a mixture or compound is calculated from the interaction coefficients of the constituent elements of the mixture or compound assuming independence of the elementary interactions. Thus

$$\frac{\mu}{\rho} = \Sigma\ \omega_i\ \frac{\mu_i}{\rho_i} \tag{2.7}$$

gives the mass attenuation coefficient of a mixture or compound, where μ_i/ρ_i is the mass attenuation coefficient of the ith element and ω_i is its fraction by weight. The assumption of independence of elementary interactions is certainly true for mixtures, and only breaks down very close to absorption edges in the case of compounds.

2.3.2 Restricted interaction coefficients

It is important in radiation dosimetry to distinguish between those parts of an interaction process which merely scatter photons and those which result in an actual transfer of photon energy to kinetic energy of charged particles.

In the photoelectric process characteristic radiation will be emitted. This will carry away a fraction of the energy $h\nu$ of the incident photon given by

$$\text{emitted fraction} = \frac{\tau_K}{\tau}\frac{\bar{E}_K}{h\nu}\omega_K + \frac{\tau_L}{\tau}\frac{\bar{E}_L}{h\nu}\omega_L = \frac{\delta}{h\nu}, \tag{2.8}$$

say, where τ_K and τ_L are the photoeffects in the K and L shells respectively, τ is the total photoeffect, ω_K and ω_L are the fluorescence yields in the K and L shells, $\bar{E}_K$ and $\bar{E}_L$ are the mean energies of fluorescence photons arising from transitions to the K and L shells, and δ is the average energy emitted as fluorescence radiation per photon absorbed.

For most biological materials this emitted fraction is quite negligible as the fluorescence yields are extremely small and $\bar{E}_K$ is only about 0.5 keV.

In the Compton effect, on the other hand, the scattered photon can carry away a very considerable fraction of the incident-photon energy, particularly at low energies. The part of the total Compton mass attenuation coefficient that actually transfers energy to electrons is σ_a/ρ.

In the pair production process an energy $2mc^2$ is emitted in the form of two annihilation photons, and the fraction converted to kinetic energy of charged particles is $1 - 2mc^2/h\nu$.

2.3.2.1 Mass energy transfer coefficient. It is convenient to derive from the total mass attenuation coefficient a coefficient that relates to the photon energy actually transferred to kinetic energy of charged particles. Such a coefficient is the mass energy transfer coefficient μ_{tr}/ρ:

$$\frac{\mu_{tr}}{\rho} = \frac{\tau}{\rho}\left(1 - \frac{\delta}{h\nu}\right) + \frac{\sigma_a}{\rho} + \frac{\kappa}{\rho}\left(1 - \frac{2mc^2}{h\nu}\right). \tag{2.9}$$

This coefficient is used to relate energy fluence of Chapter 1 to kerma of Chapter 6 (see equation (6.1)). The small nuclear photoeffect (§2.2.5) has been ignored in equation (2.9).

2.3.2.2 Mass energy absorption coefficient. All the energy transferred to kinetic energy of charged particles is not necessarily absorbed by the irradiated material as a fraction g of the charged-particle energy is converted to photon energy (bremsstrahlung, see §2.5.1). The coefficient used to derive the energy actually absorbed per unit mass in an irradiated material—the absorbed dose of Chapter 4—is the mass energy absorption coefficient μ_{en}/ρ and is given by

$$\frac{\mu_{en}}{\rho} = \frac{\mu_{tr}}{\rho}(1 - g). \tag{2.10}$$

The fraction g can be appreciable at high energies especially in materials of high atomic number, but is usually small in biological materials (see table 2.4).

2.3.3 Tabulations of data

Photon interaction coefficients for all elements over a wide energy range have been tabulated by Storm and Israel (1970). These coefficients are kept under review by J H Hubbell of the National Bureau of Standards. Hubbell *et al* (1975) have published tables of coherent and incoherent scattering coefficients and Hubbell (1977) has given values of mass attenuation and mass energy absorption coefficients for dosimetrically important low atomic number materials. The component parts of these coefficients are also given. Hubbell (1982) has extended his tabulations of mass attenuation and mass energy absorption coefficients to 40 elements and 45 mixtures and compounds over the energy range 10 keV–20 MeV. These data are based on a mixture of theoretical calculation and experimental measurements, extensive references to which are cited in the above publications. Examples of precise experimental measurements, with a discussion of the details to which attention has to be directed, have been given by Millar and Greening (1974) for low-energy photons, and by Roux (1976) for ^{60}Co γ rays.

Table 2.4 Fraction, g, of initial electron energy converted to photon (bremsstrahlung) energy on slowing to rest in various materials. From ICRU (1984c).

Initial energy/MeV	Water	Air	Al	Fe	W	Pb
0.01	0.94×10^{-4}	1.08×10^{-4}	2.13×10^{-4}	4.20×10^{-4}	1.08×10^{-3}	1.19×10^{-3}
0.10	5.84×10^{-4}	6.62×10^{-4}	1.35×10^{-3}	3.11×10^{-3}	1.03×10^{-2}	1.16×10^{-2}
1	3.58×10^{-3}	4.00×10^{-3}	7.64×10^{-3}	1.70×10^{-2}	6.03×10^{-2}	6.84×10^{-2}
10	4.07×10^{-2}	4.11×10^{-2}	7.45×10^{-2}	1.39×10^{-1}	3.01×10^{-1}	3.16×10^{-1}
100	3.19×10^{-1}	3.02×10^{-1}	4.45×10^{-1}	5.85×10^{-1}	7.53×10^{-1}	7.62×10^{-1}

Note In general the initial energy of an electron will be less than that of any photon which releases it.

2.4   Interactions of Neutrons with Matter

The interactions of neutrons with matter do not show the smooth variations with energy and atomic number that characterise most of the interactions of photons with matter. Generalisations therefore become more difficult and in the following sections the possible interactions of neutrons with matter will be indicated in broad terms and then discussion will be focused on those interactions which are most important for dosimetry in neutron therapy and radiation protection. Furthermore, whereas all photon interactions—with the exception of the relatively unimportant photonuclear effect—generate high-speed electrons in the irradiated material, neutron interactions produce a very wide range of recoil nuclei and subatomic particles, as well as generating photons which undergo all the interactions already discussed. The varied charged particles to which neutrons give rise deposit their energy in very different ways, and this has important biological consequences.

2.4.1   Elastic collisions

The simplest process is for a neutron to collide with an atomic nucleus. The neutron is deflected with some loss of energy which is transferred to the recoiling nucleus. The energy E_{tr} transferred to the nucleus of mass M_a by the neutron of mass m and energy E_n is given by

$$E_{tr} = E_n \frac{4 M_a m}{(M_a + m)^2} \cos^2 \theta, \qquad (2.11)$$

where θ is the angle of recoil in laboratory coordinates. The energy transfer is greatest when the mass of the struck nucleus is least, and this occurs for the hydrogen nucleus, the proton, of mass very close to that of the neutron. For this nucleus the cross section falls very rapidly and smoothly for the first 1 or 2 MeV and then more slowly at higher energies. This reaction is represented symbolically as ^1H (n, n) ^1H. The cross sections for other biologically important materials such as O, N and C show the same general trend but have many resonance peaks superimposed on the smooth curve.

2.4.2   Inelastic collisions

A neutron may be momentarily captured by a nucleus and then be emitted with diminished energy leaving the nucleus in an excited state. The nucleus may then return to its ground state by emission of a γ ray. One example is $^{16}O(n, n')^{16}O^*$ with the $^{16}O^*$ subsequently emitting a 6.1

MeV γ ray. The word 'inelastic' is usually reserved for reactions in which a neutron is the product particle as well as the incident particle.

2.4.3 Nonelastic collisions

If the particle resulting from the interaction is not a neutron, the descriptive term used is often 'nonelastic'. The subtleties of the English language are not always easy to follow. Thus ^{16}O (n, α) ^{13}C falls in this category. In the biologically important elements C, N and O, inelastic and nonelastic collision processes usually have energy thresholds in the range 4–12 MeV, the cross section rising sharply as the threshold energy is exceeded and reaching a fairly constant value by energies of 10–15 MeV. Some cross sections exhibit many sharp maxima (see figure 2.10).

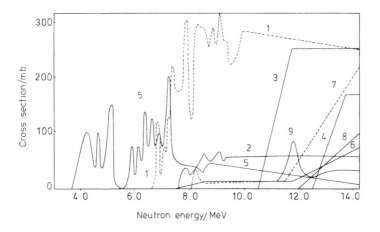

Curve		E_γ/MeV
1	$^{16}O(n, n')^{16}O*$	6.1
2	$^{16}O(n, n')^{16}O*$	7.0
3	$^{16}O(n, n')^{16}O*$	3.8
4	$^{16}O(n, n')^{16}O*$	4.8
5	$^{16}O(n, \alpha)^{13}C$	
6	$^{16}O(n, \alpha)^{13}C*$	3.1
7	$^{16}O(n, \alpha)^{13}C*$	3.8
8	$^{16}O(n, \alpha)^{13}C*$	7.0
9	$^{16}O(n, p)^{16}N$	

Figure 2.10 Inelastic and nonelastic cross sections for ^{16}O for the neutron energy range 3–14 MeV (from Auxier *et al* 1968).

2.4.4 Capture processes

Thermal neutrons, that is those in thermal equilibrium with matter, with an energy of about 0.025 eV, are captured by nuclei, with the reaction cross section frequently being inversely proportional to the neutron velocity at energies above 0.025 eV. Two important examples of this process are ^{1}H (n, γ) ^{2}H, where the γ ray has an energy of 2.2 MeV, and ^{14}N (n, p) ^{14}C which gives a 0.6 MeV proton.

2.4.5 Spallation

In this process the neutron causes fragmentation of the nucleus, several particles and nuclear fragments being ejected. The process only becomes of significance for present purposes above about 20 MeV.

2.4.6 Interaction coefficients

Whereas in the case of photons discussed earlier, greater use appears to have been made of interaction coefficients—such as the mass attenuation coefficient and the mass energy absorption coefficient—than of cross sections, the converse is the case with neutrons. However, the most useful interaction quantity in neutron dosimetry is the mass energy transfer coefficient, μ_{tr}/ρ. There are two main reasons for this. First, by multiplying this coefficient by the energy fluence (see Chapter 1) we obtain the quantity kerma (see Chapter 6) and this gives a measure of a neutron radiation field that has been used in a way analogous to that of the concept exposure (see Chapter 5) in the case of photon fields. Second, in neutron irradiations charged particle ranges are often small compared with the dimensions of volumes of interest. In such cases the kerma equals the absorbed dose which is the primary objective of radiation dosimetry. The relationship between kerma and absorbed dose is discussed in greater detail in Chapter 6.

If σ is the cross section for a particular neutron interaction with an atom of molar mass M the mass attenuation coefficient, μ/ρ, is given by (see §2.1)

$$\frac{\mu}{\rho} = N_A \frac{\sigma}{M}. \tag{2.12}$$

The mass energy transfer coefficient, μ_{tr}/ρ, for that particular interaction is then given by

$$\frac{\mu_{tr}}{\rho} = \frac{N_A \sigma}{M} \frac{\bar{E}_{tr}}{E_n}, \tag{2.13}$$

where $\bar{E}$ is the mean energy transferred to kinetic energy of charged particles by a neutron of energy E_n. The total mass energy transfer coefficient is the sum of all terms such as the right-hand side of equation (2.13) for all cross sections.

Not only are the cross sections required but it is also necessary to determine $\bar{E}_{tr}/E_n$ for each interaction. In the important case of elastic scattering this is given by

$$\frac{\bar{E}_{tr}}{E_n} = \frac{2 \, M_a m}{(M_a + m)^2}, \qquad (2.14)$$

assuming scattering to be isotropic in the centre of mass system. For other interactions such as inelastic collisions and nonelastic collisions, the matter is very much more complex and reference should be made to Bach and Caswell (1968), Caswell and Coyne (1972) or Caswell *et al* (1980) for a detailed discussion.

2.4.7 Relative importance of neutron interaction processes in H, C, N and O

The first step in the deposition of neutron energy in matter is the transfer of neutron energy to the kinetic energy of charged particles. For this

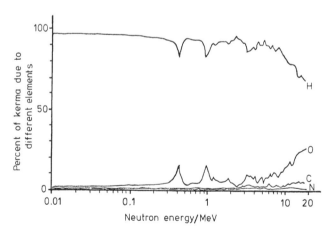

Figure 2.11 Relative contribution from interaction processes with different elements to kerma in soft tissue (R S Caswell and J J Coyne private communication).

process in soft tissue composed mainly of the elements H, C, N and O it is elastic scattering by H that is by far the most important interaction for neutrons with energies between 100 eV and 20 MeV. Indeed the ^{1}H (n, n) ^{1}H reaction accounts for some 97% of this initial energy transfer at 10 keV, is still 87% at 8 MeV and has only fallen to about 70% at 18 MeV (see figure 2.11). All this occurs despite the fact that soft tissue has only about 10% hydrogen by weight. This underlines the point that in neutron dosimetry any dosimetric material must have a hydrogen content very close to that of the tissue it is intended to simulate. The reactions coming next in importance are elastic scattering by O, C and N respectively, and it is only when neutron energies have reached about 10 MeV that inelastic and nonelastic processes equal these in importance.

The ranges of the charged particles released by neutrons even of 20 MeV will be very small. The average energy of a recoil proton from a 20 MeV neutron will be 10 MeV and such a proton will travel only 1.2 mm in water. The ranges of corresponding recoil heavy nuclei are very much smaller. Therefore the energy transferred from the initial neutrons will be absorbed very close to the point of release. This means that the processes so far discussed are the important ones for energy deposition in small volumes of tissue, but in a substantial volume of tissue the scattered neutrons undergo multiple scatterings and are eventually thermalised. At this point the capture reaction ^{14}N (n, p) ^{14}C becomes important and a neutron of very small energy causes the release of a proton of 0.62 MeV which will then transfer its energy to the tissue close to its point of origin. Furthermore, at low neutron energies the reaction ^{1}H (n, γ) ^{2}H occurs resulting in the release of a 2.2 MeV γ ray which will be added to the flux of de-excitation γ rays resulting from inelastic and nonelastic interaction processes occurring at high neutron energies. Whereas γ rays are likely to escape from small tissue volumes, they will be substantially absorbed in bigger volumes of tissue such as a man.

2.4.8 Tabulations of data

Extensive tabulations of neutron cross sections have been published by Schett *et al* (1974) and more selected data by ICRU (1977). More valuable for dosimetric purposes are the neutron kerma factors included in ICRU (1977) which give the kerma (see Chapter 6) per unit neutron fluence for 19 elements over the energy range 0.025 eV–30 MeV, and for 15 tissue compositions, compounds and mixtures over the energy range 10 eV–30 MeV. Revised values have been provided by Caswell *et al* (1980). These kerma factors are the products of the neutron mass energy transfer coefficients μ_{tr}/ρ of the materials, averaged over an energy band, and the neutron energy at the centre of that band.

2.5 Passage of Charged Particles through Matter

It has been shown in earlier sections that the interaction of photons or neutrons with matter gives rise to the release of charged particles, principally electrons and protons. In their turn these charged particles transfer their energy to the material through which they pass, and in this section consideration will be given to the manner in which they do so.

The major difference between the interactions of uncharged particles and charged particles is that in general the former suffer a small number of interactions each involving large energy losses, whereas the charged particles undergo a very large number of interactions with small energy losses. Indeed, to a first approximation, charged particles may be considered to slow down 'continuously'.

2.5.1 Radiative energy loss

Despite what has just been said, attention will first be directed to an energy loss process that results in large energy changes. On the classical electromagnetic theory a charged particle undergoing an acceleration f emits radiation at a rate proportional to f^2. For a charged particle in the field of a nucleus of atomic number Z, $f \propto Z/M$, where M is the mass of the particle, and the rate of radiation emission is therefore proportional to $(Z/M)^2$. Because of the dependence on $1/M^2$, radiative energy loss is not important for any particles heavier than an electron, and further discussion relates to that particle.

There is a rather odd convention that the radiation produced by the passage of electrons through matter is usually called bremsstrahlung if it arises in an uncontrolled manner or when the electrons are the primary interest, but is called x-radiation if it is produced deliberately in an x-ray tube. There is no fundamental difference in the interactions concerned.

Radiative energy loss increases with electron energy being very roughly proportional to energy. The rate at which a 1 MeV electron loses energy by emitting radiation in tissue is less than 1% of its total rate of energy loss, and it is not until the electron energy exceeds 100 MeV that the radiative process predominates. In a high atomic number material such as lead the situation is very different, and the radiative loss exceeds other processes even at 10 MeV.

The direction of emission of the bremsstrahlung becomes increasingly in the forward direction (i.e. the same direction as the electron producing it) as the electron energy increases. The average angle of emission is approximately equal to mc^2/E, where E is the electron energy, m its mass and c the velocity of light.

2.5.2 Collision energy loss

Far and away the most important mechanism of energy loss of charged particles is that experienced in the large number of collisions they make with atomic electrons which result in the excitation and ionisation of the material traversed.

The simple classical theory indicates that a charged particle moving past an atomic electron will impart an impulse which is proportional to the strength of the Coulomb field and the time for which it acts. The momentum gained by the atomic electron is proportional to the time of interaction, i.e. to $1/v$, where v is the velocity of the particle. Thus the energy gained by the atomic electron (or lost by the charged particle) is proportional to $1/v^2$ or to $1/E$, where E is the energy of the charged

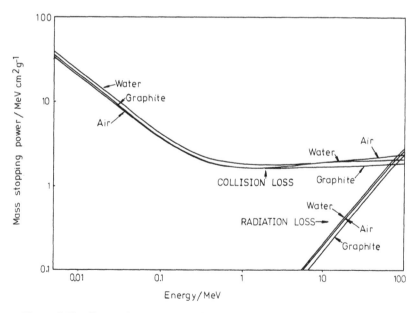

Figure 2.12   Rates of energy loss of electrons in matter (from Whyte 1959).

particle (at least under non-relativistic conditions). The energy loss dE along an increment of path dl will also be proportional to the electron density in the material traversed. At relativistic energies the particle velocity approaches constancy and the $1/E$ variation of dE/dl is modified. Indeed the contraction of the electric field makes distant collisions more probable and dE/dl increases slightly. These points are illustrated for the case of electrons in figure 2.12. The greater binding energies of high atomic number materials make excitation of inner-shell electrons less likely and in consequence dE/dl decreases slowly as Z increases (for constant electron density).

Full quantum-mechanical theory (Bethe and Ashkin 1953) predicts that dE/dl due to ionisation and excitation is proportional to the square of the charge on the particle, to the reciprocal of the square of its velocity, to the density of the medium, to Z/A of the medium and to a logarithmic function of the reciprocal of the mean ionisation potential I. The mass of the charged particle does not affect the energy loss. As Z/A and the logarithmic function of $1/I$ vary very slowly with Z, so also does dE/dl. For electrons the theory has to be modified to allow for the fact that the two particles involved in the collision are identical and the one emerging with the greater energy is regarded as the primary electron. For a detailed discussion of electron energy loss see ICRU (1984a).

2.5.3 Stopping power

The quotient dE/dl is known as the linear stopping power of a material for charged particles of energy E. The energy loss has two principal components, namely that due to collision losses and that due to radiative losses. Ignoring losses due to nuclear reactions we may write

$$S = (\mathrm{d}E/\mathrm{d}l)_{\mathrm{col}} + (\mathrm{d}E/\mathrm{d}l)_{\mathrm{rad}}, \qquad (2.15)$$

i.e.

total linear stopping power = linear collision stopping power
+ linear radiative stopping power.

It is common to express these also as mass stopping powers, and mass stopping powers are derived by dividing the linear stopping powers by the density, ρ, of the material. As the linear stopping power is proportional to density, the mass stopping power is independent of density except for the effect described in the following section. If the mass *collision* stopping power S_{col}/ρ is multiplied by the particle fluence Φ we obtain the absorbed dose of Chapter 4.

For dosimetric purposes use is sometimes made of a restricted mass stopping power, $(S/\rho)_\Delta$, in which energy losses greater than Δ are ignored. This is discussed further in Chapter 8.

2.5.4 Density effect

Distant interactions between charged particles and atomic electrons are affected by intervening atoms. These atoms will be polarised in the electric field of the charged particle and will reduce the field at a distance

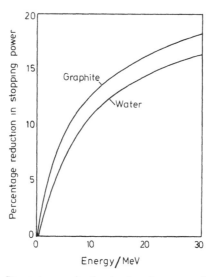

Figure 2.13 Percentage reduction in stopping power for electrons due to finite energy density effect (from Whyte 1959).

with consequent reduction in stopping power. The effect becomes more important at high energies where, as has been stated earlier, relativistic effects increase the importance of distant collisions. The effect depends on the number of atoms polarising per unit volume and thus on the density of the material—hence the name of the effect. The magnitude of the effect for electrons and its variation with energy are shown in figure 2.13. The effect is negligible for protons below about 1000 MeV and for other heavy particles. For a discussion of the density effect see Sternheimer and Peierls (1971), Berger and Seltzer (1983), and ICRU (1984c).

Normally the ratio of the mass stopping powers of two materials varies very slowly with particle energy, but, if the materials concerned are a solid and a gas, the ratio will change as relativistic particle energies are approached, due to the decrease in the mass stopping power of the solid. This is an important factor in the use of gaseous detectors to measure the radiation dose to solid or liquid materials.

2.5.5 Tabulations of data
Electron and positron stopping powers, ranges and bremsstrahlung losses in many materials have been tabulated by Berger and Seltzer (1983) and ICRU (1984c). In dosimetry it is usual to have to use ratios of mass stopping powers, and frequently the mass stopping powers are those restricted to energy losses less than Δ. Further sources of such mass stopping power ratios are given in the discussion of cavity theory in Chapter 8.

2.5.6 Linear energy transfer, L_Δ
When consideration is given to the effects of charged particles, it is often better to concentrate on the way energy is deposited in the irradiated material than on how it is lost by the particle. Some of the collisions made by a charged particle give rise to electrons which have sufficient energy to leave the main track of the charged particle and give rise to small tracks of their own. These electrons are called δ rays, and their energy is not absorbed in the immediate vicinity of the main particle track. The local energy deposition can be derived by ignoring all energy losses of the main particle which give rise to δ rays with energy exceeding some specified value Δ. Then L_Δ is the same as the linear stopping power restricted to energy losses less than Δ, i.e. $L_\Delta = (dE/dl)_\Delta$. It is usual to express Δ in eV: if $\Delta = \infty$, i.e. no restriction on the energy loss, $L_\infty = S_{col}$. For a full discussion of linear energy transfer see ICRU (1970a).

2.5.7 Mean energy expended per ion pair, W
In radiation dosimetry very extensive use is made of ionisation methods (Chapter 5 and onwards). The sensitivity of the method rests on the large number of ion pairs that can be formed in a gas by a single charged particle passing through the gas. It has been useful to introduce a quantity called the mean energy expended in a gas per ion pair formed. This is represented by the symbol W and is given by $W = E/\bar{N}$, where E is the initial kinetic energy of a charged particle and $\bar{N}$ is the mean number of ion pairs formed when the energy of the charged particle is completely

dissipated in the gas. (Electrons may dissipate some energy in bremsstrahlung production. The ions produced by the bremsstrahlung are to be included in $\bar{N}$.) The SI unit for this quantity is the joule, J, but it is also frequently expressed in electronvolts, eV.

It is fortunate for the dosimetry of electrons and photons that W for electrons is constant in most gases down to very low electron energies. The value of W for the important dosimetric material, air, has been reassessed by the ICRU (1979a). The recommended value for dry air is 33.85 ± 0.15 eV with the value for atmospheric air containing water vapour being up to 0.6% lower. The whole subject has been extensively reviewed in this same report and values of W for various charged particles and gases have been recommended. With heavy charged particles and complex gases W can vary significantly with energy and this fact adds to difficulties in the dosimetry of neutrons and heavy charged particles. Some representative values of W are given in table 2.5.

Table 2.5   Some values of average energy per ion pair (from ICRU 1979a).

Gas	Electrons $E > 10$ keV	W/eV α particles $E = 5.3$ MeV	Protons $E = 1$ MeV
CH_4	27.3	29.1	30.0
C_2H_2	25.8	27.4	
C_2H_4	25.8	27.9	
H_2	36.5	36.4	
N_2	34.8	36.4	36.5
O_2	30.8	32.2	
H_2O	29.6		
CO_2	33.0	34.2	34.5
Ar	26.4	26.3	26.5
Air	33.8	35.1	35.2
TE	29.2	31.0	30.5

Note   For uncertainties in above data and values for other gases and particle energies see ICRU (1979a).

3 Measurement of Fluence, Energy Fluence and Spectral Distributions

3.1 Principles of Measurement

Fluence and energy fluence are most readily measured in the case of collimated mono-directional beams of particles. The collimated beam is made to fall on a suitable detector which has a known efficiency of detection for the radiation concerned. The efficiency of most detectors will be a function of the particle energy, and interpretation of results requires a knowledge both of this function and the energy distribution of the particles. All this information may not be available. There are, therefore, considerable advantages in using detectors with an efficiency close to 1.0 for all particles up to the maximum energy present. Such detectors are known as total absorption devices.

3.2 Total Absorption Devices

One device of this type used for measuring the fluence of charged particles is the Faraday cup, shown schematically in figure 3.1. A beam of electrons, for example, passes through a diaphragm having a cross sectional area, a, and deposits a charge, Q, in the collecting cup. The fluence $\Phi = Q/ae$, where e is the electronic charge. The cup is placed in a high vacuum to prevent the production of ions in the space round the cup and their subsequent collection by the cup. The bottom of the cup must be thicker than the maximum range of the electrons, and is made of low atomic number material to minimise bremsstrahlung production. A layer of lead prevents such bremsstrahlung as is produced from reaching the outside of the cup and causing electron emission, and therefore charge loss. A design suitable for use with electrons up to

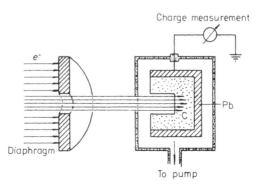

Figure 3.1 Sketch of electron fluence measurement with a Faraday cup (from ICRU 1972).

35 MeV has been described by Kretscho *et al* (1962) and extensive further references are given by Laughlin (1969) and ICRU (1972).

A total absorption device used for measuring energy fluence is the total absorption calorimeter shown schematically in figure 3.2. A beam of x-rays, for example, is defined by passing through an aperture of known area, *a*, and falls on an absorber—usually a metal of high density and high atomic number—which is of sufficient thickness to give almost total absorption of the radiation. The absorber is suspended by thin threads in a high vacuum to minimise heat losses by conduction and convection. The walls of the vacuum chamber are held at as steady a

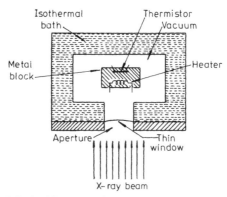

Figure 3.2 Principal features of a calorimeter for the measurement of photon energy fluence (from Whyte 1959).

temperature as possible by being surrounded by a constant temperature bath, or very good thermal insulation, or both. This stabilises radiant heat transfer between the absorber and its surroundings. A thermistor is embedded in the absorber and the change, dR_1, in its resistance on exposure to the x-rays is observed and is compared with the change of resistance, dR_2, produced by injecting a known amount of electrical energy, E, into the absorber by means of a heating coil that is embedded in it. The energy fluence, Ψ, is then given by:

$$\Psi = \frac{dR_1}{dR_2} \frac{E}{a}. \tag{3.1}$$

A total absorption calorimeter for the measurement of low-energy x-rays has been described by Greening *et al* (1968a) who also discuss the various ways in which calorimeters can be operated. A calorimeter for use with ^{60}Co radiation has been described by Genna and Laughlin (1955). Both these calorimeters have two identical absorbers suspended in the vacuum chamber, but only one absorber receives the radiation, the other absorber giving a reference against which thermal changes in the irradiated absorber can be measured. Gunn (1964, 1970, 1976) has provided an extensive review of calorimeters used as total absorption devices.

Several other types of total absorption devices have been described for use with low-energy x-rays (ICRU 1970a), including an ion chamber (Greening *et al* 1968b), a ferrous sulphate solution (Law and Redpath 1968) and silicon detectors (Greening *et al* 1969). With the ion chamber the total charge Q released by a photon beam of area a was measured. If the average energy required to produce an ion pair in the gas, W, is known, then $\Psi = QW/ea$, where e is the electronic charge. Similar treatments are possible with the ferrous sulphate solution if the radiation chemical yield of ferric ions is known, and with the silicon detectors if the average energy required to produce an electron–hole pair is known.

All these total absorption devices need corrections for any small deviations from total absorption. Losses occur by total penetration of the absorber, through backscatter and sidescatter arising from the scattering processes discussed in Chapter 2, and by fluorescence radiation produced by the photoelectric effect. These corrections have been discussed in detail by Greening and Randle (1968) for low-energy x-rays and by Pruitt and Domen (1962) for very high-energy photons.

3.3 Partial Absorption Methods

If it is not possible to achieve almost total absorption then it is probably best to go to the other extreme and use a detector that absorbs very little of the radiation as the fluence or energy fluence may then be considered to be uniform throughout the detector. Such detectors can also be used with multi-directional radiation. In practice one measures one of the dosimetric quantities, absorbed dose, D, kerma, K, or exposure, X, that are discussed in later chapters, and determines the energy fluence from the relationships between these quantities, energy fluence and the interaction coefficients of Chapter 2. For uncharged ionising particles these relationships (see Chapters 4, 5 and 6) are:

$$\Psi = \frac{D}{(\mu_{en}/\rho)_m} \quad \text{under conditions of charged particle equilibrium}$$

(3.2)

$$\Psi = \frac{K}{(\mu_{tr}/\rho)_m}. \tag{3.3}$$

Additionally for photons only

$$\Psi = \frac{X\, W_{air}/e}{(\mu_{en}/\rho)_{air}}, \tag{3.4}$$

where $(\mu_{en}/\rho)_m$ and $(\mu_{tr}/\rho)_m$ are respectively the mass energy absorption and the mass energy transfer coefficients of the detector material, $(\mu_{en}/\rho)_{air}$ is the mass energy absorption coefficient of air, W_{air} is the mean energy required to produce an ion pair in air, and e is the electronic charge. The relationships are given here as they provide a means of determining energy fluence, but they cannot be properly appreciated until the dosimetric quantities are understood.

The interaction coefficients all vary with particle energy and so a knowledge of the latter is necessary if appropriate values of the coefficient are to be chosen. If the particles have a range of energies the coefficients used must be weighted according to the spectral distribution of energy fluence with respect to energy. In all cases the fluence, Φ, can be obtained from the relationship between fluence and energy fluence, namely $\Phi = \Psi/\bar{E}$, where $\bar{E}$ is the mean particle energy weighted by fluence.

It is more difficult to measure neutron fluence than it is to measure the fluence of photons or charged particles. Thermal neutrons can be measured by absorbing them in gold foils and then measuring the activity that has been induced in the foil. Fast neutron fluence can be determined using detectors which measure the recoil protons produced by the neutrons in hydrogenous materials. These and many other techniques have been reviewed by ICRU (1969a).

3.4 Determination of Spectral Distributions

The following sections will indicate briefly the methods that can be used for the determination of spectral distributions and will give references to publications where further details and references may be found.

3.4.1 Photons
X-rays (bremsstrahlung) are produced as a result of radiative energy loss by electrons slowing down on passage through a material (§2.5.1). These x-rays have a continuous range of energies from that of the most energetic electron downwards. Other x-rays are produced by radiative transitions of electrons within atoms (§2.2.1.4) and these x-rays occur at discrete energies characteristic of the atom from which they come. The inner shell vacancies from which these characteristic x-rays arise may have been produced either directly by electron bombardment or indirectly by photoelectric self-absorption of the bremsstrahlung.

3.4.1.1 Low-energy photons. Spectral distributions of low-energy x-rays have been reviewed by Greening (1972). The earliest method used, and the one giving greatest resolving power, was crystal diffraction. The method requires the application of several corrections to the raw data, and modern measurements including consideration of these corrections have been reported by Gilfrich and Birks (1968). Kramers (1923) put forward a theory from which spectral distributions could be calculated and this has shown fair agreement with various experimenters over the years once allowance is made for attenuation within the x-ray tube target and other filtering materials (Unsworth and Greening 1970, Soole 1971). Sunderaraman *et al* (1973) have used Monte Carlo techniques to calculate low-energy spectra and find fair agreement with Kramers' theory. Reiss and Steinle (1973) have carried out many Monte Carlo calculations of spectra in the x-ray diagnostic energy range.

Silberstein (1932) showed how attenuation curves in various materials could be converted to spectral distributions. The method was extended and generalised by Greening (1950) and has found further application by Burke and Pettit (1960), Saylor (1969) and Twidell (1970). Attenuation analysis continues to be used (Baird 1981, Archer and Wagner 1982) at least in part because of its undemanding requirements for apparatus. It is particularly useful when the radiation is too intense to be measured by counting techniques. It should be mentioned here that the early results of the attenuation analysis method and of crystal diffraction have usually been expressed as distributions in wavelength, λ, rather than in energy, E. Care is needed in converting these to distributions in energy. An ordinate y_1 in a distribution against wavelength has to be multiplied by $d\lambda/dE$ to convert it to the corresponding ordinate y_2 in a distribution against photon energy, since $y_1 d\lambda = y_2 dE$. As λ is proportional to E^{-1}, $d\lambda/dE$ is proportional to E^{-2}.

Spectral distributions can also be obtained by using detectors which have energy discrimination, such as a sodium iodide crystal–photomultiplier combination, a gas proportional counter, or lithium drifted germanium or silicon detectors. The resolution of these detectors, expressed as the full width at half height of the distribution of pulses produced by mono-energetic photons, improves in absolute terms (i.e. the width decreases on an energy scale) as the photon energy decreases, but worsens in relative terms (i.e. the width becomes a greater percentage of the photon energy). It is approximately true to say that the width expressed as a percentage of the photon energy, E, is proportional to $E^{-1/2}$, and pulse size analysis methods are thus seen to deteriorate in resolution as the photon energies decrease.

The resolution is also affected by the average energy which has to be deposited in the detector in order to produce a single electrical charge. This energy varies by roughly an order of magnitude between each of the detectors mentioned above, being very roughly 300 eV for a sodium iodide crystal–photomultiplier, 30 eV for a gas proportional counter and 3 eV for the germanium or silicon detectors.

For work with germanium and silicon detectors see Drexler and Perzl (1967, 1968) (figure 3.3), Drexler and Gossrau (1968), Gossrau and Drexler (1971), Mika and Reiss (1968, 1969a,b), Cameron and Ridley (1970) and Walter (1970). Birch *et al* (1979) have provided a valuable catalogue of spectra between 30 and 140 kV based on calculations supported by measurements with a germanium detector. Gas proportional counters have been used by Clark and Gros (1968, 1969), Cairns *et al* (1969), Hink *et al* (1970) and Unsworth and Greening (1970), while scintillation spectrometers were employed for spectral distribution

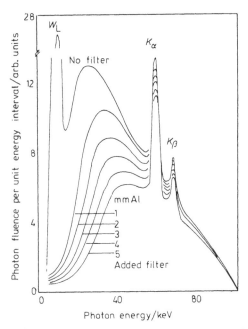

Figure 3.3 Spectral distributions of x-rays from a tube with 1 mm Be inherent filtration operated at 100 kV and with various filters (measured after passage through 2 m of air) (from ICRU 1970b).

measurements of low-energy x-rays by Epp and Weiss (1966), Peaple and Burt (1969) and Unsworth and Greening (1970).

The measurement of a photon spectral distribution is a substantial undertaking. In consequence use is often made of a spectrum measured by other people for what is supposed to be similar equipment operated under the same conditions. In fact the equipment may not be quite the same and the operating conditions may not be accurately known (for example, maximum energy and energy distribution of electrons producing the photons; target material, angle and condition; filtration). There is often much to be said for using an approximate representation of the photon beam based on a few simple attenuation measurements made with it. This approach has been discussed in detail and useful data tabulated by Greening (1963).

3.4.1.2 High-energy photons. A major review of bremsstrahlung production by multi-million eV electrons has been given by Koch and Motz

(1959). More recent experimental work on accelerators used for medical purposes has been reported by Bentley *et al* (1967), O'Dell *et al* (1968), Rawlinson and Johns (1973), Levy *et al* (1974), and Huang *et al* (1981, 1982). Spectral distributions at these high energies are substantially different for thick and thin targets in the accelerators.

For γ-ray sources the energies of the photons and their relative abundance can be obtained from tabulations such as those of Lederer *et al* (1967, 1978) or Dillman and Von der Lage (1975).

3.4.2 Electrons
The measurement of electron spectra has been briefly reviewed by ICRU (1972, 1984a). Spectra of electrons with energies below 0.1 MeV can be measured with an electrostatic analyser, semiconductor detectors (silicon and germanium) can be used up to a few MeV, and at higher energies scintillation spectrometers (Feist *et al* 1968) or magnetic spectrometers (Siegbahn 1965) are employed.

The spectra of β-ray emitters are usually well established. An example

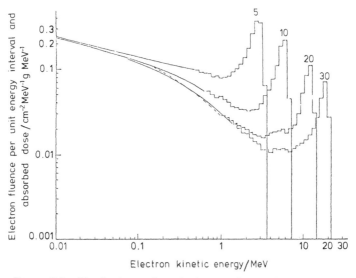

Figure 3.4 Distributions of total electron fluence in energy, normalised to absorbed dose in water, for primary electron beams of 5, 10, 20 and 30 MeV at depths in water such that the primary electrons have travelled 0.4 to 0.5 of their range (to convert to units of cm^{-2} MeV^{-1} Gy^{-1} multiply ordinates by 6.24×10^9) (from Nahum 1975).

is given in figure 1.1, and many β emitters will give spectra of similar shape. The spectra of a large number of β emitters of clinical interest have been given by Cross *et al* (1983). The electrons produced in some types of accelerator, e.g. betatron and synchrotron, are nearly mono-energetic immediately before issuing from the accelerator as the latter can be, in effect, a spectrometer itself. After passing through the window of the vacuum chamber of the accelerator the electrons will have a slightly lower energy and a somewhat greater energy spread. The approximate energy at this stage can be determined by:

(a) observing the accelerator setting at which a nuclear reaction of known threshold energy is induced (ICRU 1972, 1984a);

(b) measuring the range of the electrons in appropriate absorbers (Pohlit 1969); or

(c) observing the threshold of production of Cerenkov radiation in a gas as its pressure, and therefore its refractive index, is varied (Liesem 1976).

Once an electron beam enters an irradiated medium the spectral

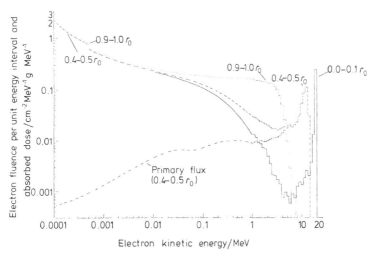

Figure 3.5 Distributions of total electron fluence in energy, normalised to absorbed dose in water, for a primary electron beam of 20 MeV at depths in water such that the primary electrons have travelled 0 to 0.1, 0.4 to 0.5 and 0.9 to 1.0 of their range, r_0 (to convert to units of cm^{-2} MeV^{-1} Gy^{-1} multiply ordinates by 6.24×10^9) (from Nahum 1975).

distribution is substantially altered by attenuation, scatter, and the production of secondary and higher-order electrons. Although experimental measurements have been made, the electron spectra used in dosimetry are usually derived from theoretical calculations using electron transport theory (e.g. Kessaris 1970) or the Monte Carlo method (e.g. Berger *et al* 1975, Nahum 1975, 1978). Examples are given in figures 3.4 and 3.5. The Monte Carlo technique, as applied in radiation physics, has been reviewed by Raeside (1976).

3.4.3 Neutrons

The distribution in energy of neutron fluence or fluence rate can be determined by a number of techniques including

(a) time of flight measurements,

(b) observation of proton recoils in gas proportional counters, solid state detectors, solid and liquid scintillators or photographic emulsions,

(c) threshold detectors, and

(d) fission chambers.

These and other techniques have been reviewed by ICRU (1969a) and many examples of spectra given. The neutron sources of most interest for medical applications are ^{252}Cf, d–T generators, and cyclotrons producing

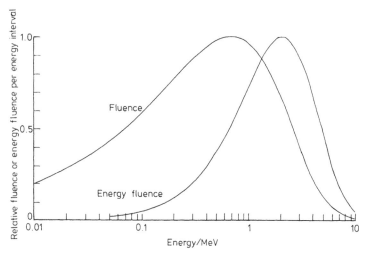

Figure 3.6   Differential distribution in energy of fluence and energy fluence of neutrons from ^{252}Cf.

deuterons or protons for bombardment of beryllium targets. Spectra of [252]Cf have been reported by Werle and Bluhm (1972), Knitter *et al* (1973) and Ing and Cross (1975). Werle and Bluhm (1972) state that for [252]Cf $\Phi_E \propto E^{1/2} \exp(-E/T)$ where, if energy is expressed in MeV, $T = 1.42\,\text{MeV}$ (see figure 3.6). Hannan *et al* (1973) report measurements of d–T neutrons from a source with various collimators both in air and at depths in tissue-equivalent liquid, while Ing and Cross (1975) derive neutron spectra for [252]Cf and d–T neutrons in a cylindrical phantom by Monte Carlo methods. Mijnheer *et al* (1981) have measured d–T neutron spectra in a water phantom using activation and fission detectors. Parnell (1972) used a recoil proton spectrometer to measure the neutrons arising from the [9]Be (d, n) reaction at 16 MeV (see figure 3.7) while Meulders

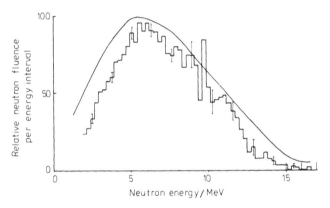

Figure 3.7 Comparison of neutron spectra produced by bombarding thick beryllium and carbon targets with 16 MeV deuterons. Full curve, beryllium; histogram, carbon (from Parnell 1972).

et al (1975) measured neutron spectra resulting from the same reaction at 16, 33 and 50 MeV using the time of flight method. Further work with deuterons and/or protons on [9]Be has been reported by Lone *et al* (1977, 1981), Johnsen (1977), Heintz *et al* (1977), Amols *et al* (1977), Madey *et al* (1977), Waterman *et al* (1979), Graves *et al* (1979) and Ullmann *et al* (1981). Neutron spectra at a depth in water- or tissue-equivalent phantoms arising from the [9]Be (d, n) reaction have been published by Bonnett and Parnell (1976), Mountford *et al* (1976), and Bonnett and Parnell (1982).

3.5   Weighted Values of Interaction Coefficients and Stopping Powers

In later chapters there will be frequent use of interaction coefficients such as μ_{en}/ρ and μ_{tr}/ρ for photons and neutrons and of stopping powers and stopping power ratios for charged particles. If the photons, neutrons and charged particles concerned have a range of energies, as is usually the case, then the interaction coefficients and stopping powers will need to be weighted according to the energy distribution of the appropriate property of the particles. For simplicity later equations are written for mono-energetic particles, but the need for weighting should always be borne in mind.

4 Direct Measurement of Absorbed Dose

4.1 History of Absorbed Dose

Many of the laws of science—for example, all the 'conservation' laws, the laws of thermodynamics, Newton's first and third laws of motion, Lenz's law in electromagnetism and Le Chatelier's principle in physical chemistry—are special and more precise statements of the everyday phrase, 'You do not get something for nothing.' The same applies to effects produced by ionising radiations and it has therefore always been obvious that these effects can only be brought about if the radiation transfers some of its energy to the material in which the effect is produced. The process of energy absorption is important even if it is later followed by complex physical, chemical or biological processes. Thus there has long been a desire to determine the energy absorbed by irradiated materials as a step in obtaining a quantitative correlation between the radiations and the effects they produce. The heating effects of x-rays on metal foils were measured as early as 1897, and Curie and Laborde (1904) used a calorimetric method to determine the rate of energy release by radium. By 1913 Christen was advocating the use of a quantity he called 'dose' and which he defined as the radiant energy absorbed per unit volume. (An account in English (Christen 1914) may be more readily available.) He recognised that measurements would probably have to be made of the number of ions produced in a specified volume of dry air under standard conditions of temperature and pressure, but thought it would be better to calculate the energy required to set free this number of ions so that 'the unit of dose would be expressed by ergs per cubic centimetre'. He was ahead of his time. Zimmer (1938) and Gray and Read (1939) were faced with the problem of measuring neutrons. As these particles could not be measured in terms of the quantity *exposure* which had by that time been brought into use for x-ray and γ-ray measurements (see §5.1), they introduced an 'energy unit' for the purpose. Like Christen's unit it was expressed in energy per unit volume and was calculated relative to the energy absorbed per unit volume of water exposed to one roentgen (§5.1) of γ rays. But again like Christen they had to measure it in terms of ionisation in gases because, as will be seen later, there are severe practical difficulties in the direct measurement of ionising radiations by the heating effect they produce. It was not until after its meeting in 1950 that the ICRU (1951) felt it prudent to say 'For the correlation of the dose of any ionising radiation with its biological or related effects the ICRU recommends that the dose be expressed in terms of the quantity of energy absorbed per unit mass (ergs per gram) of irradiated material at the point of interest.' Even then they went on to say 'Inasmuch as calorimetric methods are not usually practicable, ionisation methods are generally employed.' The ICRU (1954) gave this quantity the name *absorbed dose* and said 'The rad is the unit of absorbed dose and it is 100 ergs per gram.' At its 1956 meeting the ICRU (1957) said the 'absorbed dose of any ionising radiation is the energy imparted to matter by ionising particles per unit mass of irradiated material at the point of interest.' The 1959 report (ICRU 1961) gave guidance on the term 'energy imparted' and the 1962 report (ICRU 1962) said 'The absorbed dose (D) is the quotient of ΔE_D by Δm where ΔE_D is the energy imparted by ionising radiation to the matter in a volume element, Δm is the mass of matter in that volume element.' The Δs were introduced to indicate that Δm was sufficiently small to define the point of measurement but sufficiently large for ΔE_D to be made up of so many energy deposition events that a good average value was obtained. Absorbed dose was therefore a macroscopic quantity needing to be averaged over a volume, in the same way as other macroscopic quantities such as density have to be averaged. The ICRU (1971a) changed absorbed dose to a quantity defined at a point by a differential quotient $d\bar{\varepsilon}/dm$, where $d\bar{\varepsilon}$ was the *mean* energy imparted, with the meaning given in the present day definition below.

Currently (ICRU 1980) the important dosimetric quantity absorbed dose, D, is defined by $D = d\bar{\varepsilon}/dm$ where $d\bar{\varepsilon}$ is the mean energy imparted by ionising radiation to material of mass dm. Now it is unusual in defining a physical quantity by a differential quotient to specify a mean value, since mean (or expectation) values are normally understood to apply. The reason for its use here lies in the nature of the quantity ε (see next section), and its importance in some aspects of dosimetry. The energy imparted, ε, by ionising radiation to the matter in a volume is

$$\varepsilon = \Sigma R_{\text{in}} - \Sigma R_{\text{out}} + \Sigma Q,$$

where ΣR_{in} is the radiant energy incident on the volume, i.e. the sum of the energies (excluding rest energies) of all those charged and uncharged ionising particles which enter the volume; ΣR_{out} is the radiant energy emerging from the volume, i.e. the sum of the energies (excluding rest energies) of all those charged and uncharged ionising particles which leave the volume; and ΣQ is the sum of all changes (decreases: positive sign, increases: negative sign) of the rest mass energy of nuclei and elementary particles which occur in the volume. In most instances ε is given simply by the difference between the sums of the energies of those particles entering and those leaving the volume. If, however, the volume contains a radioactive source that emits a particle that escapes from the volume, the energy of this particle would increase ΣR_{out} and lead to an underestimate of ε were it not for the ΣQ term. Corresponding considerations apply if one of the incoming particles enters a nucleus and changes its rest mass.

4.2 Stochastic and Non-stochastic Quantities

Ionising radiations interact with matter by one or more of the discrete processes discussed in Chapter 2. Furthermore, the emission from radiation sources will vary with direction and time. In consequence there may frequently be substantial variations in ε that have to be considered in detail. It is, therefore, a stochastic quantity; its values vary discontinuously in space and time, and one does not refer to its rate of change. Measurements of ε if repeated often enough can establish an experimental distribution, and the average of these experimental values of ε will give an estimate of $\bar{\varepsilon}$ which is a non-stochastic quantity. Thus absorbed dose, namely $d\bar{\varepsilon}/dm$, is a non-stochastic quantity, is defined at a point and is in general a continuous function of space and time having a gradient and a rate.

The stochastic quantity corresponding to absorbed dose is *specific energy*, z, defined as $z = \varepsilon/m$. If the energy fluence, or the interaction coefficients, or the mass m are small, z can fluctuate widely in repeated measurements, but, as with ε, enough of these measurements will establish a mean $\bar{z}$. This is a non-stochastic quantity, and the limit of mean specific energy, $\bar{z}$, as the mass, m, tends to zero is the absorbed dose

$$D = \lim_{m \to 0} \bar{z}.$$

From this definition and the previous discussion it will be seen that strictly it is always the quantity z that is measured, not D, which is something of theoretical abstraction. In most instances z is sufficiently large, and the individual energy deposition events sufficiently numerous, for the statistical fluctuations in z on repeated measurements to be small or even negligible compared with other uncertainties in the measurements. However, when m is very small the number of energy deposition events can be small and variations in z significant. We are then in the realm of microdosimetry which will not be discussed further in this book. Reference should be made to ICRU (1984b) or to the proceedings of the Microdosimetry Conferences organised by Euratom every two years (Euratom 1983).

4.3   Units for Absorbed Dose

The original special unit of absorbed dose, the rad, is still in use unchanged after a quarter of a century. It was originally expressed in terms of ergs, namely $100 \, erg \, g^{-1}$, but with the growing use of SI units was expressed in terms of those units as $10^{-2} \, J \, kg^{-1}$. The rad was at first reserved for use only with the quantity absorbed dose, but the ICRU (1971a) later extended its use to other radiation quantities having the same dimensions, namely specific energy (imparted), z, kerma, K, and absorbed dose index, D_I (see §12.7). With the increasing use of the SI system of units the ICRU, after widespread sounding of opinion, asked the General Conference on Weights and Measures to sanction a special name, gray, symbol Gy, for the SI unit $J \, kg^{-1}$ when used with ionising radiation quantities. It was thought that a special name would be of particular value to radiation users. This was agreed in 1975, and the gray is now the recommended unit for absorbed dose, specific energy, kerma and absorbed dose index. The old special unit, rad, can be used temporarily with the SI system but is recommended to be phased out by about 1985. Some countries have already abandoned its use.

As there are four quantities that can be expressed in grays it is necessary to state what quantity is involved when a number of grays is mentioned unless this is clear from the context. Furthermore, as absorbed dose, kerma, and specific energy can apply to any material, a statement of absorbed dose, etc is incomplete without an indication of the material concerned.

4.4 Absorbed Dose Calorimeters

In principle absorbed dose can be measured directly by observing the heating effect produced by ionising radiation in the material of interest. In practice there are many problems. Stahel (1929) made an attempt to measure the energy absorbed per unit volume of water exposed to high-energy photons by the combined use of a calorimeter and a liquid filled ionisation chamber, but the results were rather inaccurate. Hochanadel and Ghormley (1953) used calorimetry to determine the energy absorption per unit mass of a few cm^3 of the ferrous sulphate dosimeter solution (§11.3) in a glass vessel, but it was not until 1956 that a calorimeter was used successfully to measure the absorbed dose in a small mass of material surrounded by a much larger mass of the same material (Genna and Laughlin 1956). This is called a homogeneous calorimeter and a modern design is shown schematically in figure 4.1 (Domen and Lamperti 1974).

A mass of the material in which the absorbed dose is to be determined—called the core or thermal element—is surrounded by a mass of similar material—usually called the jacket—from which it is thermally insulated by a narrow evacuated gap. In many designs the jacket is, in its turn, surrounded by and isolated from a further mass of the material, usually called the mantle or shield. All this is embedded in, but isolated from, the main bulk of the material of the calorimeter. Depending upon the mode of operation of the calorimeter, all or some of the core, jacket, and shield will have temperature sensing devices (thermistors or thermocouples) and electrical heaters embedded in them. During irradiation the calorimeter of figure 4.1 operates in a quasi-adiabatic mode as the temperature of the jacket closely follows that of the core. During calibration a known amount of electrical energy is fed to a heater in the core. As the jacket is designed to have the same mass as the core, the core's heat loss is allowed for by adding the temperature rise of the jacket to that of the core. The heat loss of the jacket to the shield is a second-order effect.

Let us suppose that the irradiated material in a calorimeter is water. This has a specific thermal capacity of about 4200 J kg^{-1}°C^{-1}. If an absorbed dose of 2 Gy (200 rad) is delivered—a typical daily absorbed dose in radiotherapy—the resulting temperature rise will be $2/4200 \doteqdot 0.0005$ °C. If this is to be measured with 1% accuracy it will be necessary for the temperature measuring system to be able to detect a temperature change of a few millionths of a °C! If graphite were the

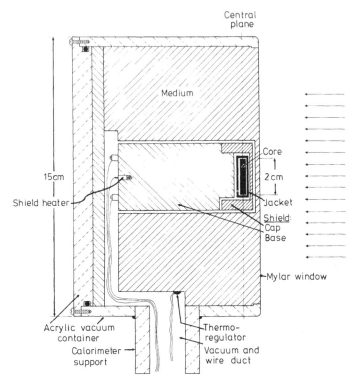

Figure 4.1 Side view cross section of the NBS portable calorimeter (from Domen and Lamperti 1974).

irradiated material, as has been the case with several calorimeters including that of figure 4.1, the temperature rise would be about six times as
great, but its accurate measurement would still constitute a problem.
The small amount of energy to be measured means that even very small
heat exchanges between the isolated mass and its surroundings can have
a significant effect on the measurement. If the thermal element is made
small, in order the better to define the point of measurement, the surface
to volume ratio increases and accentuates the effects of these heat
exchanges. In practice there are additional reasons for not reducing the
mass of the thermal element. As has been said it needs to contain a
thermistor and a heating element. These will be made of materials
which differ from that of the thermal element and will therefore absorb

radiation in a different manner. The mass of the thermal element must therefore be large compared with that of the inserted materials if these are not seriously to disturb the measurement.

A further complication is that not all the heating effect in the thermal element is necessarily due to the energy imparted, as given in the definition of absorbed dose. The radiation can induce endothermic or exothermic chemical changes in the thermal element or energy can be stored in a crystal lattice. The material of this element has to be chosen with care so that it has either a negligible or an accurately known 'heat defect'. The basic equation relating the absorbed dose, D, at a point in the thermal element with the energy, E_H, released as heat and the energy, E_S, stored chemically or physically in the element is

$$D = \frac{dE}{dm} = \frac{dE_H + dE_S}{dm}.$$

E_S is called the heat defect or thermal defect. It is positive if energy is stored (e.g. in a crystal lattice) or if an endothermic reaction takes place. It is negative for an exothermic reaction. For a discussion with further references see Bewley *et al* (1974). Simple materials such as metals or graphite are least likely to cause problems in this respect. However, if the calorimetric material is different from that in which the absorbed dose is required, there will be a problem in transferring the absorbed dose in the calorimeter material to that in the material of interest. First, different materials may affect the radiation field in different ways producing different radiation fluences in the two materials. Second, the materials will absorb different fractions of the radiation energy depending upon such of their properties as energy absorption coefficients and stopping powers. This latter problem is common to all dosimetry where the sensitive dosimetric material and the material in which the absorbed dose is required differ, and is discussed in later chapters.

Laughlin and Genna (1966) have provided a thorough examination of the problem of radiation calorimetry with a section devoted to absorbed dose determination. Gunn (1964, 1970, 1976) has published reviews with very extensive lists of references to calorimetry. Further discussions are provided by ICRU (1969b) and Radak and Marković (1970), while McDonald *et al* (1976) describe a modern portable calorimeter. An ingenious design of small graphite calorimeter that can be immersed in a tank of water and used for the direct measurement of absorbed dose has been published by Rao and Naik (1980).

Absorbed dose calorimeters such as that shown in figure 4.1 are difficult to construct as they have several parts surrounded by other parts from which they have to be thermally insulated, but all parts need electrical leads for temperature sensors and heating elements. Their manufacture is not to be lightly undertaken. However, Domen (1980) has described an absorbed dose water calorimeter of delightful simplicity. Essentially it consists of a 0.25 mm diameter thermistor sandwiched between two very thin films of polyethylene held horizontally in a plastic frame immersed in an insulated but unregulated tank of water. The water is irradiated from the top so that the thermal gradient discourages convection which is further inhibited by the polyethylene film. The low thermal diffusivity of water has permitted the measurement of absorbed dose rates in water of about 4 Gy min^{-1} (400 rad min^{-1}) with a precision of 0.5%. Measurements with this device after development in its construction and mode of operation (Domen 1982, 1983a, 1983b) have given absorbed dose values with ^{60}Co radiation about 3.5% higher than those measured with a graphite calorimeter. This difference is currently attributed to a heat defect in water, and future use of water calorimeters will require further investigation of this heat defect.

Work is under way to establish calorimetric standards of absorbed dose in several national standards laboratories. It is perhaps a measure of the difficulty of carrying out calorimetry with the accuracy required by standards laboratories that work from the NBS was first published by Petree and Ward (1962) and work from NPL by Kemp *et al* (1971), and yet by 1980 few calibrations against absorbed dose standards had been made. This is not to say that considerable progress has not been made. More recent work from standards laboratories has been reported by Guiho *et al* (1978) who compared the calorimetric absorbed dose standards of France and the USA using ^{60}Co γ rays and found excellent agreement. These calorimetric standards have also been compared with the ionometric standard of the BIPM using ^{60}Co radiation (Allisy 1979). The observed differences of about 0.3% are well within the estimated systematic errors of the various methods.

The fundamental problems with calorimetric methods of dosimetry are their lack of sensitivity and the difficulty of ensuring low heat exchange between the thermal element and its surroundings in what are often constrained experimental conditions. Under special circumstances these problems can disappear. Thus when there is a need to measure intense beams of pulsed radiation, calorimetry comes into its own. Just when other methods of measurement are running into absorbed dose rate problems users of calorimetry find that there is plenty of energy to absorb and the irradiation time is so short that there is little opportunity for heat transfer to take place. An example of such usage of calorimetry has been given by Epp *et al* (1975) and the whole subject is reviewed by ICRU (1982).

Nevertheless calorimetric methods of dosimetry are used in only a small proportion of measurements and it is to a much more sensitive method that we must turn in the next chapter.

5 Exposure and Its Measurement

5.1 History of Exposure

It should come as no surprise that the so-called ionising radiations—the radiations with which this book is concerned—have very frequently been measured by the ionisation they produce. Early investigations of x-rays (Thomson and Rutherford 1896) and electrons were made by observing the ionisation they produced in gases, and the Curies' measurements of their radioactive ores during their refinement for the final separation of radium were also made by ionisation methods. These methods were used because of their sensitivity. A single α particle dissipating its energy in a gas might produce of the order of 10^5 ion pairs. A γ ray could release a Compton recoil electron that would produce of the order of 10^4 ion pairs in a gas, and even a relatively low-energy photon, such as occurs in diagnostic radiology, could give rise to about 10^3 ion pairs. The substantial electric charge released by a single ionising particle, coupled with the sensitivity of even early electroscopes and electrometers, made ionisation methods very attractive for the quantitative measurement of x-rays, γ rays, electrons and other charged particles.

It was much easier to collect ions released in gases than it was to collect ions released in liquids or solids. It was also convenient to use air as the gas in an ionisation chamber as it would be there anyhow unless special measures were taken to replace it by another gas. Accordingly, as early as 1908 Villard (1908) was suggesting as a unit for a rather vague 'quantity' of x-rays the amount that would produce 1 esu of electric charge in $1 \, \text{cm}^3$ of air at $0 \, °C$ and $760 \, \text{mmHg}$ pressure. There were problems in defining the conditions under which this ionisation was to be produced. It was recognised that some of the ionisation arose from interactions of the x-rays with the air in the ion chamber, and other ionisation was produced by the passage through the air of electrons knocked out of the wall of the ion chamber—the so-called 'wall effect'. An additional reason for the choice of air as the standard dosimetric material was that the main pressure for accurate measurement of x-rays was coming from their medical use. Here there was a desire to correlate the x-rays quantitatively with the biological effects they produced. It was therefore sensible to choose as the standard dosimetric material one that interacted with x-rays in the same way as soft tissue. As air had much the same average atomic number as did soft tissue, air was a good choice.

The First International Congress of Radiology was held in 1925 and created the International x-ray Unit Committee which in 1928 became the International Commission on Radiological Units (ICRU). (Nowadays this is the International Commission on Radiation Units and Measurements, but it retains the same abbreviation.) In 1928 the ICRU defined a unit, called the roentgen, symbol r, to be used with x-radiation saying that 'This international unit be the quantity of x-radiation which, when the secondary electrons are fully utilised and the wall effect of the chamber is avoided, produces in one cubic centimetre of atmospheric air at 0 °C and 76 cm mercury pressure, such a degree of conductivity that one electrostatic unit of charge is measured at saturation current.' It should be noted that there was no special name for the 'quantity' concerned, the word being used synonymously with 'amount', and the nature of the quantity itself was vague. This point was not to be resolved until 1956.

In 1937 the ICRU changed the definition of the roentgen as follows (ICRU 1938). 'The international unit of quantity or dose of x-rays shall be called the 'roentgen' and shall be designated by the symbol 'r' . . . The roentgen shall be the quantity of x- or γ radiation such that the associated corpuscular emission per 0.001293 g air produces in air ions carrying 1 esu of quantity of electricity of either sign.' The apparently arbitrary mass of air of 0.001293 g is the mass of air in 1 cm^3 at 0 °C and 76 cm Hg pressure. Note that γ rays could now also be measured in this unit and that the 'quantity' measured was even more confused. This problem of the quantity being measured was resolved at the 1956 meeting of ICRU when it defined a quantity 'exposure dose'. The definition (ICRU 1957) ran as follows: 'Exposure dose of x- or γ radiation at a certain place is a measure of the radiation that is based upon its ability to produce ionisation.' This was followed by: 'One roentgen is an exposure dose of x- or γ radiation such that . . .' and the rest of the definition remained as in 1937. In 1962 the ICRU dropped the word 'dose' from 'exposure dose' to help to avoid confusion with the quantity absorbed dose which was by then firmly in existence. Also, in keeping with the convention that the symbol for a unit named after a person should have a capital letter, the roentgen was given the symbol R instead of r. The ICRU (1968) modified the definition of exposure so that it was 'the quotient of ΔQ by Δm where ΔQ is the sum of the electrical charges on all the ions of one sign produced in air when all the electrons (negatrons and positrons) liberated by photons in a volume element of air whose mass is Δm are completely stopped in air. $X = \Delta Q/\Delta m$.' (For meaning of Δ see p48.) Furthermore the roentgen (R) = 2.58 × 10^{-4} C kg^{-1}. The apparently arbitrary number 2.58 × 10^{-4} had arisen through changes over the years (i) from a volume to a mass of air and (ii) from an esu of charge to an SI unit of charge. In 1971 (ICRU 1971a) the ΔQ and Δm were changed to true differentials dQ and dm so that exposure could be said to exist at a point and to have a spatial gradient.

The latest definition of exposure (ICRU 1980) says: 'The exposure, X, is the quotient of dQ by dm where the value of dQ is the absolute value of the total charge of the ions of one sign produced in air when all the electrons (negatrons and positrons) liberated by photons in air of mass dm are completely stopped in air. $X = dQ/dm$.' The unit is now to be the SI unit C kg^{-1} but the special unit of exposure, roentgen (R), can be used temporarily, and 1 R = 2.58 × 10^{-4} C kg^{-1} (exactly).

5.2   Requirement of Definition of Exposure

All these definitions from 1928 to the present day require that the only material to enter into the chain of interactions is air. Firstly, the photons are to interact with a defined mass of air. In so doing they will produce electrons by the photoelectric effect and Compton effect and both electrons and positrons by the pair production process. All these secondary charged particles must then travel through air until their energy is dissipated and the ions of one sign that are produced must be measured.

The secondary electrons will not lose all their energy by collision processes, but will lose a small amount by radiative losses, i.e. bremsstrahlung production. Any ionisation caused by subsequent reabsorption of the bremsstrahlung is *not* to be included in the ionisation specified in the definitions. This means that the photon interaction coefficient related to the definition of exposure must exclude any bremsstrahlung production component and therefore is the mass energy absorption coefficient, μ_{en}/ρ. The charged particle energy released per unit mass of air which is subsequently reabsorbed by the air is then $\Psi(\mu_{en}/\rho)_{air}$, where Ψ is the energy fluence. If the average energy required to produce an ion pair in air is W_{air}, the number of ion pairs per unit mass will be $\Psi(\mu_{en}/\rho)_{air}/W_{air}$ and the charge, Q, produced per unit mass will be $\Psi(\mu_{en}/\rho)_{air}e/W_{air}$, where e is the electronic charge. But exposure $X = dQ/dm$, i.e. charge per unit mass in the above phraseology. Therefore

$$X = \Psi \left(\frac{\mu_{en}}{\rho}\right)_{air} \frac{e}{W_{air}}. \tag{5.1}$$

No one has yet carried out an experiment in which the precise requirements of the definitions of exposure are met. It would not seem possible to irradiate a defined mass of air and yet avoid irradiating a surrounding mass of air that will be required for the complete stopping of the charged particles released in the defined mass. Recourse has to be made to the concept of charged particle equilibrium.

5.3 Charged Particle Equilibrium

In figure 5.1 suppose a large volume V_1 of air to be uniformly irradiated throughout by photons. This could be achieved if there were negligible attenuation of the photons in the air volume. In a small volume such as V_2 a number of electrons will be released and these will travel partly through V_2 and partly through the surrounding V_1 producing ions. V_1 is sufficiently large that electrons from V_2 are completely stopped in it. But every volume such as V_2 in V_1 will undergo the same interactions with

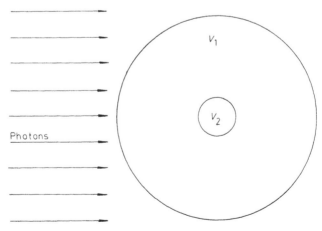

Figure 5.1 Principle of electronic equilibrium.

the photons and the number of electrons released and their distributions in energy and direction will be the same. They will also produce the corresponding ions along their tracks. From symmetry there will be no build up of the secondary electrons in any particular volume such as V_2. There will be the same number, energy and direction of the electrons entering V_2 as of those leaving it. We thus have a condition of charged particle equilibrium. Furthermore, the ionisation produced in V_2 by all the electrons passing through it from the rest of V_1 will equal the total ionisation produced in the rest of V_1 by the electrons released by the photons in V_2. A version of this condition of charged particle equilibrium is used in the device—the 'free-air' chamber—that is employed as a standard of exposure.

It has been assumed above that the air is uniformly irradiated. In practice photons are attenuated to some extent on passage through any medium. As a result complete charged particle equilibrium is rarely achieved. This point is discussed further in §6.4.

5.4 The Free-air Chamber

A free-air chamber is shown schematically in figure 5.2. A mono-directional beam of x-rays passes through a defining aperture of accurately known area, A, enters a metal box and passes out through a hole on the far side of the box without striking anything in the box other than the air it contains. This fulfils the requirement that the x-rays interact

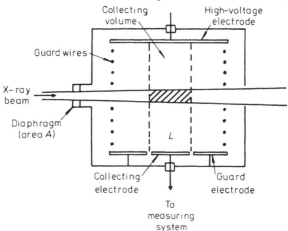

Figure 5.2 Free-air chamber (from Whyte 1959).

only with air. The separation of the electrodes in the chamber and its other dimensions are such that electrons released in the shaded volume lose all their energy before they can reach the electrodes or chamber walls. This ensures that the electrons are completely stopped by air. A high potential difference maintained between the high-voltage electrode and the collecting electrode of length L sweeps the ions of one sign produced between the broken lines to the collecting electrode. A system of guard wires, to which graded electrical potentials are applied, together with the guard electrodes, ensures that the field lines between the collecting electrode and the high-voltage electrode are straight and perpendicular to those electrodes. The ions are therefore collected from a defined volume. It remains to determine the volume, and thence the mass, of air in which the electrons producing this ionisation are effectively produced using a special case of the principle of charged particle equilibrium mentioned earlier. This volume will be shown to be that corresponding to the shaded region in figure 5.2.

Figure 5.3 reproduces the x-ray beam and broken lines of figure 5.2 and adds the tracks of three electrons ejected from air lying in the path of the x-ray beam. Electron track A originates in the shaded volume of figure 5.2. The ionisation in that part of it lying in the shaded volume of figure 5.3 will be collected, but that lying outside the shaded volume will be lost. However, this loss will be exactly compensated by the ionisation produced along those parts of the electron tracks B and C lying in the shaded volume. (This concept of electronic equilibrium depends on the photon attenuation being negligible over distances equal to the ranges of the electrons. This is substantially true for the photon beams for

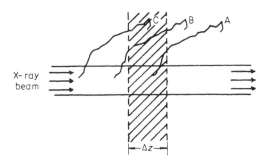

Figure 5.3 Electronic equilibrium in free-air chamber (from Whyte 1959).

which free-air chambers are used as exposure standards, i.e. for x-rays generated at potential differences up to about 300 kV.)

It is usual to state the exposure, X, at the position of the diaphragm. The reason is as follows. The source of x-rays is not a point, so the shaded volume of figure 5.2 has a penumbra round its curved surface and its cross-sectional area is ill defined. However, at the diaphragm the beam cross section, A, is known. The ionisation produced by the electrons ejected from the air in any length L of the diverging pencil of x-rays will be independent of the position within the column if there is negligible attenuation of x-rays in the air. The measured ionisation is therefore the same as would be produced in an air column of length L and area A at the diaphragm. The column's volume is then AL and the mass of air in it is $AL\rho$, where ρ is the air density. Then $X = Q/\rho AL$, where Q is the charge flowing to the measuring system.

Several corrections, usually small, have to be made to allow *inter alia* for (i) any attenuation of the x-rays in the air between the diaphragm and the centre of the collecting electrode, (ii) any lack of saturation in the ionisation current, (iii) any water vapour present and (iv) any ionisation resulting from bremsstrahlung, scattered radiation or radiation leaking into the chamber through its walls.

Wyckoff and Attix (1957) have made a detailed examination of the design of free-air chambers and the many corrections that have to be applied to measurements with them. Kemp (1977) has reviewed more recent developments in the subject. Sectional views of the NBS and NPL exposure standards for use up to about 300 kV are given by Aston and Attix (1956). Free-air chambers for use up to about 50 kV have been described by Greening (1960), Lamperti and Wyckoff (1965) and Boutillon *et al* (1969). The last reference also describes intercomparisons of the standards at BIPM, NBS and NRC (Canada).

5.5 Exposure Measurement with Calibrated Cavity Chamber

A free-air chamber for the measurement of x-rays generated at 300 kV would have an electrode separation of about 30 cm. The overall lateral dimensions would be about 45 cm and the length perhaps 60 cm. It is therefore a bulky device and needs a substantial high-voltage supply, and is thus inconvenient to use except under restricted laboratory conditions. Most importantly it only accepts radiation from a very small solid angle. For convenience in day to day measurements in the field, and particularly when multi-directional radiation is concerned, it is necessary to use a different device.

In figure 5.1 suppose it were possible to compress the air in volume V_1 by a factor of 1000 without compressing the air in volume V_2. All the electrons in V_1 would have their ranges reduced by a factor of 1000, but the number, energy and direction of those crossing V_2 would be unchanged. A formal proof of this has been given by Fano (1954) (see Chapter 8). We would then have a compact device which gave electronic equilibrium, and if we could measure the ionic charge produced within it and the mass of air it contained, their quotient would give the exposure. This is a hypothetical situation but it can be approximated to in practice. The 'condensed air' round V_2 can be replaced by a conducting solid with an atomic number close to that of air. This solid is made in two parts separated by an insulator also with atomic number close to that of air (figure 5.4). If a potential difference is established between the two parts of the air-equivalent conducting wall, the ionisation can be collected and led away to a measuring system. In practice a geometrical distortion to the shape of figure 5.5 usually takes place and leads to the schematic design of one form of what is called a cavity ionisation chamber. Such a device can be calibrated against a free-air chamber by placing the free-air chamber in an x-ray beam and measuring the exposure, and then moving the free-air chamber to one side and placing the cavity chamber at the point previously occupied by the diaphragm of the free-air chamber. This establishes a calibration factor which removes the need to know the mass of gas in the cavity or the sensitivity of the measuring system. (*Changes* in the mass of gas due to changes of temperature or pressure must, however, be allowed for.) Furthermore, because the

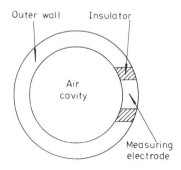

Figure 5.4 Principle of the cavity chamber.

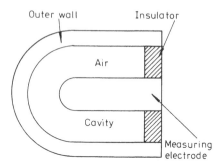

Figure 5.5   Practical shape of the cavity chamber.

cavity chamber wall has an atomic number close to that of air, the calibration factor will only vary slowly as the energy of the x-ray beam is changed and in consequence the x-ray spectrum does not need to be accurately known. The vast majority of exposure measurements have been made with a device of this kind.

5.6   Free-air Chambers for High-energy Radiation

For x-rays generated above 300 kV, or for γ rays of corresponding energy, the electrode separation required in free-air chambers increases rapidly as the photon energy rises. Kaye and Binks (1937) built a free-air chamber with an electrode separation of three metres to measure radium γ rays, while Taylor and Singer (1940) tackled the problem by constructing a chamber that operated at a pressure of several atmospheres. Wyckoff (1960) has described a more modern high-pressure chamber. These heroic methods presented substantial difficulties, and an alternative approach to the provision of exposure standards for higher-energy radiations has had to be sought. If with the cavity ionisation chamber described earlier, it were possible to determine (i) the mass of gas in the cavity, and (ii) the extent to which the cavity wall deviated from 'air equivalence', such a device could become an exposure standard for use above 300 kV. The first point presents no fundamental problems, but discussion of the second will be left to Chapter 8 where cavity theory is considered.

6 The Concept of Kerma and Its Relationships

6.1 Introduction and Definition

Following a suggestion by Roesch (1958) the ICRU (1962) defined a quantity it called kerma. This name is an acronym for Kinetic Energy Released per unit Mass, the final 'a' being added principally to avoid confusion with 'kern' in spoken German. This quantity was introduced to emphasise the two stage process that takes place when indirectly ionising particles such as photons or neutrons impart energy to matter. In the first stage the uncharged particles transfer energy to the kinetic energy of charged particles. In the second stage those charged particles impart energy (in the sense used in the definition of absorbed dose) to matter. The quantity kerma is useful in considering the first of these two processes. It was first defined (ICRU 1962) as 'the quotient of ΔE_k by Δm, where ΔE_k is the sum of the initial kinetic energies of all the charged particles liberated by indirectly ionising particles in a volume element of the specified material and Δm is the mass of matter in the volume element.' Δ has the meaning indicated in the discussion of the early definitions of absorbed dose in Chapter 4. The ICRU (1971a) dropped the Δs and used differentials thus making kerma a point function, in the same way and at the same time as it modified the definition of absorbed dose. In addition, the symbol E_k was changed to E_{tr} thus making the definition $K = dE_{tr}/dm$.

Currently (ICRU 1980) the kerma, K, is defined as 'the quotient of dE_{tr} by dm where dE_{tr} is the sum of the initial kinetic energies of all the charged ionising particles liberated by uncharged ionising particles in a material of mass dm,' i.e. $K = dE_{tr}/dm$. The unit of kerma is J kg^{-1} which, as with absorbed dose and specific energy, discussed in Chapter 4, is given the special name gray, Gy. The special unit of kerma is the rad, 10^{-2} J kg^{-1}, which can be used temporarily but is due to be phased out by about 1985. A statement of kerma is incomplete without a reference to the material concerned.

6.2   Kerma and Energy Fluence

Kerma, K, is related to the energy fluence, Ψ, by the mass energy transfer coefficient, μ_{tr}/ρ, discussed in Chapter 2:

$$K = \Psi \frac{\mu_{tr}}{\rho}. \tag{6.1}$$

Since dE_{tr} is the sum of the initial kinetic energies of the charged particles liberated by the uncharged ionising particles, it includes the energy these charged particles later radiate in bremsstrahlung.

6.3   Kerma in Air and Exposure

It was seen in Chapter 5 that in the measurement of exposure any ionisation resulting from bremsstrahlung was not to be included. Were it not for this, exposure would be the ionisation equivalent of air kerma, that is the ionisation per unit mass of air resulting from the energy transfer per unit mass of air. From equation (5.1) exposure, X, is given by

$$X = \Psi \left(\frac{\mu_{en}}{\rho}\right)_{air} \frac{e}{W_{air}}.$$

From equation (6.1) air kerma, K_{air}, is

$$K_{air} = \Psi \left(\frac{\mu_{tr}}{\rho}\right)_{air}. \tag{6.2}$$

Therefore

$$K_{air} = \frac{X\,W_{air}}{e} \left(\frac{\mu_{tr}/\rho}{\mu_{en}/\rho}\right)_{air}, \tag{6.3}$$

$$= \frac{X\,W_{air}/e}{1 - g}, \tag{6.4}$$

where g is the fraction of electron energy lost in bremsstrahlung production (see Chapter 2). This fraction, g, is only significant at high energies, and only amounts to about 0.3% for electrons released in air by ^{60}Co radiation. It falls rapidly with decreasing photon energy as, first, the fraction of the photon energy transferred to electron energy falls and, second, the radiation yield of that electron energy is roughly proportional to the initial electron energy. Thus as W_{air}/e is accurately known (see §2.5.7), the free-air chamber of Chapter 5 can be regarded as a standard of air kerma as well as a standard of exposure.

6.4 Kerma and Absorbed Dose

The relationship between kerma and absorbed dose is more subtle. The energy transfer of kerma takes place at a point, but the subsequent imparting of energy to matter which gives rise to the absorbed dose is spread over distances determined by the ranges of the charged particles.

6.4.1 Without bremsstrahlung

Consideration will first be given to the relationship between absorbed dose and kerma when bremsstrahlung production is ignored. The matter will be discussed in terms of photons and the electrons to which they give rise, but it is qualitatively the same for neutrons and the recoil protons or heavy recoil nuclei which they produce.

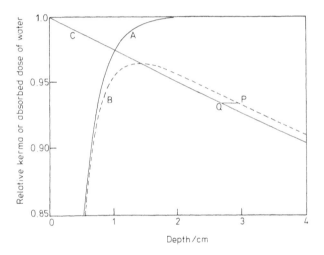

Figure 6.1 Absorbed dose and kerma to water irradiated by parallel beam of photons from 6 MeV linear accelerator, assuming no bremsstrahlung. A, absorbed dose if no attenuation of primary photons; B, actual absorbed dose; C, kerma.

Figure 6.1 shows the kerma and absorbed dose resulting when a beam of high-energy photons falls on a material from a vacuum (the particular case illustrated is for photons from a 6 MeV linear accelerator irradiating water). The kerma is highest at the point where the photons first fall on the water and then falls off with depth due to photon attenuation processes. The absorbed dose in a thin surface layer will be small as few electrons will be released in it and be able to deposit energy in it. Indeed most of the energy will be deposited by electrons scattered back from the underlying material. In an adjacent deeper layer the fluence of electrons and the absorbed dose will be greater as, in addition to the few electrons released in the layer itself, there will be others coming from the overlying layer. As the depth increases so also will the electron fluence and the absorbed dose until the increase of electron fluence produced by an additional layer is balanced by the attenuation of photons 'upstream'. This attenuation of photons brings about a decrease of electrons at downstream points. Thereafter there is a continuing fall of electron fluence and absorbed dose because of this attenuation of photons. In figure 6.1 curve A shows the build-up of absorbed dose that would have occurred in the absence of photon attenuation, curve B shows the actual absorbed dose and curve C the steady decrease of kerma. The main features of these curves and some other useful results are demonstrated by the following simple mathematical treatment.

Suppose a parallel beam of high-energy photons of energy fluence Ψ_0 falls normally on the surface of some medium. If the beam is broad the energy fluence within the medium near the central axis will show no lateral variation and the energy fluence will decrease down the central axis with an effective linear attenuation coefficient μ_p, say. The Compton recoil electrons (or any electron–positron pairs) will tend to be projected in the same direction as the photons producing them. In any event the gradient of the electron energy fluence will be zero at right angles to the central axis of the beam and the electron energy fluence will be assumed to fall off axially with an effective linear attenuation coefficient μ_e. This adoption of a constant effective attenuation coefficient for electrons with its consequence of exponential attenuation is an approximation, but it is in reasonable agreement with experiment. It permits a simple instructive treatment that is mathematically the same as that used in calculating the initial growth and subsequent decay of a short-lived radioactive daughter of a long-lived parent radionuclide.

As the linear attenuation coefficient of the photons is μ_p the photon energy fluence at a depth x will be $\Psi_0 \exp(-\mu_p x)$ and this will release an electron energy fluence of $\Psi_0 \mu_{tr} [\exp(-\mu_p x)] \, dx$ in a thickness dx of medium where μ_{tr} is the energy transfer coefficient of the medium for the photons. In travelling to a greater depth x_1 the electron energy fluence will be attenuated by the factor $\exp[-\mu_e(x_1-x)]$, and at x_1 a fraction $\mu_e dx$ will be absorbed in a thickness dx. The absorbed dose in the volume of unit area, thickness dx and density ρ at x_1 is given by

$$D_{x_1} = \Psi_0 \frac{\mu_{tr}}{\rho} \mu_e \int_0^{x_1} \exp(-\mu_p x) \exp[-\mu_e(x_1 - x)] \, dx, \qquad (6.5)$$

whence

$$D_{x_1} = \Psi_0 \frac{\mu_{tr} \mu_e}{\rho(\mu_e - \mu_p)} [\exp(-\mu_p x_1) - \exp(-\mu_e x_1)]. \qquad (6.6)$$

But the kerma at x_1, K_{x_1} will be

$$K_{x_1} = \Psi_0 \frac{\mu_{tr}}{\rho} \exp(-\mu_p x), \qquad (6.7)$$

and in general differs from D_{x_1}.

From equations (6.6) and (6.7) it may be shown that $K_{x_1} = D_{x_1}$ when

$$x_1 = \frac{\ln(\mu_e/\mu_p)}{\mu_e - \mu_p}. \qquad (6.8)$$

Also by differentiating equation (6.6) it may be shown that D_{x_1} is a maximum for the same value of x_1 as is given by equation (6.8). Thus the kerma curve of figure 6.1 intersects the absorbed dose curve at its maximum.

Dividing equation (6.6) by (6.7) it follows that

$$\frac{D_x}{K_x} = \frac{\mu_e}{\mu_e - \mu_p} \{1 - \exp[-(\mu_e - \mu_p)x]\}, \qquad (6.9)$$

and when x is large enough to make the exponential term negligible this reduces to

$$\frac{D_x}{K_x} = \frac{\mu_e}{\mu_e - \mu_p}, \qquad (6.10)$$

and the absorbed dose and kerma bear a constant ratio to one another with the absorbed dose being the greater.

Furthermore from equations (6.6) and (6.7) it can be shown that on the falling part of the absorbed dose and kerma curves $D_x = K_{x-d}$ where $d = \mu_e^{-1}$ if $\mu_e \gg \mu_p$. The absorbed dose at P in figure 6.1 is equal to the kerma at Q a distance μ_e^{-1} 'upstream'. In exponential attenuation the attenuation coefficient is the reciprocal of the mean range of the attenuated particles. Therefore $d = \mu_e^{-1}$ is the mean range of the electrons. This result is the same as that arrived at by more general methods by

Roesch (1958) in his original paper on kerma. He has given a more formal mathematical treatment later (Roesch 1968).

The absorbed dose that would be obtained with negligible photon attenuation is given by equation (6.6) with $\mu_p = 0$. This condition corresponds to curve A of figure 6.1.

It may also be shown that at the point where the kerma and absorbed dose curves intersect the primary radiation has been attenuated by the factor $(\mu_p/\mu_e) \ln(\mu_e/\mu_p)$.

6.4.2 With bremsstrahlung

Consideration will now be given to the case in which bremsstrahlung production cannot be ignored. This will only occur with secondary electrons arising from photon irradiation, as bremsstrahlung production by the protons released by neutron irradiation is negligible even when the protons have energies of 1000 MeV.

In deriving equation (6.5) it was stated that of the electron energy fluence reaching the depth x_1 a fraction $\mu_e dx$ was absorbed in a thickness dx. Now μ_e is the *attenuation* coefficient of the electrons. It is the *absorption* coefficient, to be used in calculating absorbed dose, only if none of the attenuation is due to radiative energy loss (bremsstrahlung

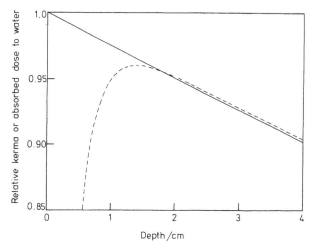

Figure 6.2 Absorbed dose and kerma to water irradiated by parallel beam of photons from 6 MeV linear accelerator, taking bremsstrahlung into account. Broken line, absorbed dose; full line, kerma.

production). If there is bremsstrahlung production μ_c' must be substituted for μ_c in the numerators outside the integral and square brackets of equations (6.5) and (6.6) respectively, where μ_c' is that part of μ_c that arises from collision energy loss. This means that the absorbed dose is everywhere reduced by the factor μ_c'/μ_c. The situation with bremsstrahlung production is illustrated in figure 6.2. Proportionality between kerma and absorbed dose still occurs in the region where it did without bremsstrahlung but now the absorbed dose curve has been lowered all along its length, and the two curves no longer intersect at the peak of the absorbed dose curve.

For this case allowance for bremsstrahlung has brought the absorbed dose and kerma curves almost into coincidence. The coincidence is even closer for 2 MV, ^{60}Co and 4.5 MeV linear-accelerator radiations. Thus for this range of photon energies kerma and absorbed dose are very nearly equal in water on the falling part of the absorbed dose curve.

Attix (1968) was the first to draw attention to the lowering effect bremsstrahlung had on the absorbed dose curve. He has developed the point further (Attix 1979) showing the effects of bremsstrahlung on absorbed dose in aluminium. As bremsstrahlung production is roughly proportional to the atomic number of the medium, it is greater in aluminium than in the water considered above and the absorbed dose curve is brought down below the kerma curve.

6.4.3 Charged particle equilibrium

At the depth where curve A of figure 6.1 has built up to its maximum the electron fluence is a maximum and a stable electron spectrum has been established. Charged particle equilibrium exists and continues at greater depths. The situation represented by curve A will only arise if the photons suffer negligible attenuation in distances equal to the maximum ranges of the electrons they eject. (These distances will be about five times μ_e^{-1}, the mean electron range, used in the mathematical treatment of §6.4.1.) Some relevant data are given in table 6.1, and it will be seen that with low-energy photons and even quite energetic neutrons full charged particle build-up can occur with negligible attenuation of uncharged particles. In such circumstances the absorbed dose equals the kerma. Strictly it should be added that this is so only if bremsstrahlung production is negligible, but it will be negligible in low atomic number materials under the conditions required for negligible photon attenuation, and in any case for neutrons.

The situation is different if the attenuation of uncharged particles is

Table 6.1 Approximate thickness of water required to establish charged particle quasi-equilibrium, and attenuation of uncharged particle in that thickness.

Maximum energy† of accelerator/MeV	Approximate thickness of water for equilibrium/mm	Approximate attenuation/%
0.3	0.1	0.03
0.6	0.4	0.1
1	0.8	0.3
2	2.5	0.8
4	8	2
6	15	4
8	25	6
10	30	7
15	50	9
20	60	11
30	80	13
Neutron energy‡/MeV		
0.1	0.001	0.005
1	0.02	0.04
10	1.3	0.5
30	10	1.5

† The mean photon energy would be about 0.37 times this energy.
‡ This is the mean neutron energy not an accelerator energy.

not negligible in the build-up region. Thus at the point where the absorbed dose and kerma curves of figure 6.1 intersect there is a local equilibrium between the energy transferred to kinetic energy of charged particles and the energy absorbed from charged particles. The term transient equilibrium is sometimes applied to the region in which the absorbed dose and kerma curves are almost parallel, but a better term for this region is quasi-equilibrium or perhaps proportional equilibrium to reflect the condition of equation (6.10) which shows that absorbed dose and kerma are proportional to one another here.

Even at the point of local equilibrium between energy absorbed and energy imparted there will not quite be true charged particle equilibrium in all respects (i.e. in number, energy and direction) as it is not until depths are reached at which the 'unattenuated' curve A of figure 6.1

reaches its maximum that a stable electron spectrum is established. At high photon energies complete electronic equilibrium is rare, and therefore whenever a statement is made that something is true 'under conditions of electronic equilibrium' it is probable that in practice a correction—hopefully small—will need to be applied to allow for lack of this equilibrium.

6.4.4 Near interface between two media

The discussion of figure 6.1 concerned the relationship between absorbed dose and kerma within a single medium. It is useful to extend this discussion to the relationship that exists near the boundary between two media. Figure 6.3 illustrates the situation in two media in which the energy fluence of uncharged particles is uniform. The kerma in the two media will in general be different because the media will have different mass energy transfer coefficients, and the ratio of the kermas on the two sides of the interface between the media is the same as the ratio of these coefficients. The absorbed dose in medium 2 is explained in the same way as that of curve A in the single medium of figure 6.1. In medium 1 the absorbed dose and kerma will be equal until points are reached which are within range of charged particles ejected back or scattered

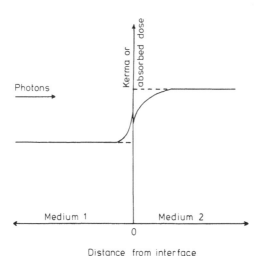

Figure 6.3 Relationship between kerma and absorbed dose at the interface between two media. Broken line, kerma; full line, absorbed dose.

back from medium 2. At such points the absorbed dose in medium 1 will increase, and will be at a maximum at the interface with medium 2. At this interface the energy fluence of charged particles will be the same in both media, but the absorbed doses will not be the same. This is because the mass stopping powers of the media for the charged particles are not the same, and the absorbed doses on the two sides of the interface will be in the same ratio as the mass stopping powers. If figure 6.3 is understood many of the problems of dosimetry will be resolved. Further consideration will be given to absorbed dose in the vicinity of interfaces when dosimeter probes are discussed in Chapter 8.

7 Determination of Absorbed Dose via Exposure or Air Kerma

7.1 Absorbed Dose in Air

It was indicated in Chapter 4 that there are as yet no national standards of absorbed dose in routine use, although appropriate calorimeters are being developed. The only dosimetric quantities for which national standards at present exist are exposure and air kerma. Determinations of absorbed dose have therefore to be related to these two standards. It is usual to calibrate secondary standards—usually cavity ionisation chambers—against these national standards and then to use the secondary standards to measure exposure directly or to calibrate tertiary field instruments. These exposure measurements have then to be converted to absorbed dose measurements as described below.

It has been shown in Chapter 5 that exposure X is dQ/dm where dQ is the charge produced in air by the secondary electrons ejected by photons from a mass dm of air. If the mean energy required to produce an ion pair in air is W_{air}, the energy imparted to the air by the electrons released from the mass dm is

$$dQ\frac{W_{air}}{e} = X\,dm\,\frac{W_{air}}{e},$$

but the energy is not deposited in the identical mass of air that the secondary electrons came from. However, if we establish electronic equilibrium as discussed in conjunction with figure 5.1, the energy deposited in each mass dm of air will be the same, and in particular the energy deposited in the mass dm from which the secondary electrons were ejected will also be $X\,dm\,W_{air}/e$. Then the energy per unit mass of air is $X\,W_{air}/e$ and this is, of course, the absorbed dose in the air. Therefore:

$$D_{air} = X\frac{W_{air}}{e}. \tag{7.1}$$

Thus, *under conditions of electronic equilibrium*, the absorbed dose in air is a factor W_{air}/e times the exposure. Furthermore, and most fortunately for radiation dosimetry, except at *very* low energies which are not usually of consequence in dosimetry, W_{air}/e is constant for electrons in air. Thus absorbed dose to air is directly proportional to exposure at all energies if electronic equilibrium applies.

(As pointed out in §6.4.3, true electronic equilibrium rarely exists and we have to settle for a quasi-equilibrium. In such circumstances there is some attenuation of radiation between the point at which electrons are released and the point at which they deposit their energy. This entails a small correction which amounts to about 0.5% for ^{60}Co radiation and diminishes rapidly as the photon energy decreases.)

7.2　Absorbed Dose in Other Materials

It is also simple to derive absorbed dose in materials other than air from an exposure measurement, again providing electronic equilibrium obtains. The energy absorbed per unit mass of various materials subjected to the same energy fluence will be proportional to the mass energy absorption coefficients, μ_{en}/ρ, of those materials. But from equation (7.1) the absorbed dose in air is $X W_{air}/e$, so that the absorbed dose, D_m, in some other material is given by

$$D_m = X \frac{W_{air}}{e} \frac{(\mu_{en}/\rho)_m}{(\mu_{en}/\rho)_{air}}. \tag{7.2}$$

For materials with atomic number close to that of air the factor $(\mu_{en}/\rho)_m/(\mu_{en}/\rho)_{air}$ varies only slowly with photon energy. The values of this ratio for water and air are shown as a function of photon energy in figure 7.1. It will be seen that the overall variation is only about 2% between 100 keV and 10 MeV, and only 10% for the three decades of photon energy from 10 keV to 10 MeV. In consequence the photon energy does not have to be known with great accuracy in order to determine the absorbed dose in water from an exposure measurement.

For materials of higher atomic number, such as bone, or those with a high hydrogen content, such as fat, this fortunate state of affairs no longer applies, and the ratio of the mass energy absorption coefficients of bone and fat to those of air are illustrated in figure 7.2. In fact, in the case of bone the problems are even greater than figure 7.2 might indicate. Bone is not a homogeneous material as it has various soft tissue elements

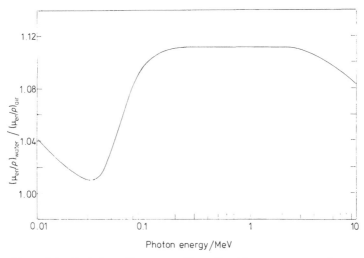

Figure 7.1　Variation with energy of the ratio of the mass energy absorption coefficients of water and air.

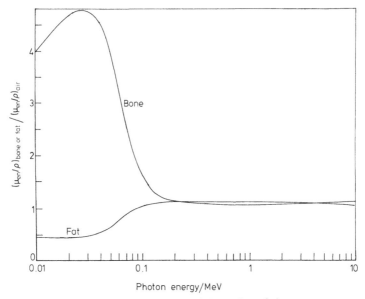

Figure 7.2　Variation with energy of the ratios of the mass energy absorption coefficients of bone and fat to that of air.

within an inorganic matrix of higher atomic number. The dimensions of these soft tissue elements are often of the same order as the ranges of the electrons released in the bone during photon irradiation, and we are faced with the problem of dosimetry near interfaces that was illustrated in a simplified form in figure 6.3. For a discussion of bone dosimetry with further references see King and Spiers (1985).

7.3 Factors Converting Exposure to Absorbed Dose

If in equation (7.2) we use the special radiation units, rad for absorbed dose and roentgen for exposure, and take W_{air} to be 33.85 eV (i.e. $W_{air}/e = 33.85$ J C^{-1}) we find (see Appendix)

$$D_m = 0.873 \frac{(\mu_{en}/\rho)_m}{(\mu_{en}/\rho)_{air}} X \frac{rad}{R}. \tag{7.3}$$

The product $0.873 \, (\mu_{en}/\rho)_m/(\mu_{en}/\rho)_{air}$ has frequently been represented by the symbol f. From figure 7.1 it is seen that the mass energy absorption coefficient ratio in the case of water is approximately 1.1 over a wide energy range. The numerical value of the f factor thus becomes 0.96. The very close numerical agreement between an exposure expressed in R and the corresponding absorbed dose to water expressed in rad, has been a matter of convenience, although some would argue that it has stopped people thinking carefully enough about the difference between the quantities exposure and absorbed dose.

If, however, in equation (7.2) we use SI units, expressing the absorbed dose in Gy and the exposure in C kg^{-1} we have (see Appendix)

$$D_m = 33.85 \frac{(\mu_{en}/\rho)_m}{(\mu_{en}/\rho)_{air}} X \frac{Gy}{C \, kg^{-1}}. \tag{7.4}$$

In the case of water the new numerical value of the f factor becomes about 37 over a wide energy range and the simple approximate numerical equivalence of the quantities absorbed dose in water and exposure has disappeared.

7.4 Calibrations in Terms of Air Kerma

The position can be largely restored by providing a calibration in terms of air kerma instead of exposure. Air kerma is the energy equivalent of the air ionisation in the definition of exposure, with a correction—usually very small—for bremsstrahlung production. The energy per unit mass equivalent of exposure X is $X\,W_{air}/e$ and the allowance for bremsstrahlung production is made by using the ratio of the mass energy transfer and mass energy absorption coefficients for air. Thus

$$K_{air} = X\,\frac{W_{air}}{e}\,\frac{(\mu_{tr}/\rho)_{air}}{(\mu_{en}/\rho)_{air}}. \tag{7.5}$$

Substituting the value of $X\,W_{air}/e$ from this equation in equation (7.2) we obtain, under conditions of electronic equilibrium,

$$D_m = K_{air}\,\frac{(\mu_{en}/\rho)_m}{(\mu_{tr}/\rho)_{air}}. \tag{7.6}$$

Both absorbed dose and kerma can be expressed in the same units (either, *both* in rad or *both* in Gy) and in the case of water the ratio of the interaction coefficients is 1.08 ± 0.03 between 0.1 and 10 MeV, giving fairly close numerical agreement between absorbed dose in water and air kerma. Because of this approximate agreement, and because air kerma is measured in the same units as absorbed dose whereas exposure is measured in $C\,kg^{-1}$, it is highly probable that the use of the quantity exposure will die out and that air kerma will take its place.

7.5 Calibrations in Terms of Absorbed Dose to Water

It might be supposed that if the absorbed dose in a standard material such as water can be derived from exposure through equation (7.2), it should be possible to provide an instrument that had been calibrated against an exposure standard with a calibration in terms of absorbed dose to water. It is possible but it presents problems. The reason is that whereas both exposure and kerma relate conceptually to initial interactions with infinitesimal isolated masses of material, absorbed dose is influenced by charged particles coming from a finite volume of material surrounding the point of interest. The experimental conditions employed have to be specified, and require the application of a number of small correction factors. These could, in principle, be determined by standards laboratories for their specified conditions of measurement or could, as in the past, be left to the user to determine for his own experimental situation. Views differ as to which is the better approach. However, several national standards laboratories have decided to mount a joint programme to assess these problems, and if they can ultimately agree common conversion and correction factors, it will do much to unify the determination of absorbed dose throughout the world.

7.6   High-energy Calibrations

National standards laboratories do not have exposure standards for photons more energetic than x-rays produced at 2 MV or γ rays from ^{60}Co. Their standards at these energies are based on cavity ionisation chambers. Many users of ionising radiations need to determine absorbed dose in materials irradiated by radiations with energies greater than— often much greater than—those for which an exposure calibration can be provided. The derivation of absorbed dose at these high energies is related to the lower energy exposure calibration by means of cavity theory. In consequence, both the exposure standards themselves and the absorbed dose determinations are left to the next chapter where cavity theory is discussed.

8 Determination of Absorbed Dose and Exposure from Cavity Theory

8.1 The Bragg–Gray Cavity Theory

In order to measure the absorbed dose in a medium it is necessary to introduce a radiation sensitive device into it. As in general the device will not be of the same material as the medium, the radiation detector constitutes a foreign body in the medium. Throughout the time modern dosimetry theory has developed, gas ionisation techniques have predominated in precise work. The radiation sensitive device has usually been a gas-filled cavity and the associated theory has been called 'cavity' theory. In principle the theory applies to any foreign body, solid, liquid or gas, introduced into the medium.

The foundations of modern cavity theory were laid by Gray (1929, 1936) although Bragg (1912) had earlier presented a qualitative discussion. Gray considered a medium uniformly irradiated by photons and of sufficient dimensions that electronic equilibrium was established at a place within it at which a small gas-filled cavity was introduced. He showed (Gray 1929) that the introduction of the cavity did not modify the number, energy or direction of the electrons crossing the cavity surface providing that scattering of electrons in the cavity could be neglected. He felt this was justifiable if the cavity were sufficiently small, i.e. that the electrons lost only a small fraction of their energy in crossing the cavity. If both the solid medium and the gas have the same number and energy of electrons passing through them, the ratio of the electron energy lost per unit mass in the two materials will be the same as the ratio, s_{mg}, of the mass stopping powers, $(S/\rho)_m$ and $(S/\rho)_g$, of the medium and the gas for the electrons concerned. This follows directly from the definition of mass stopping power given in Chapter 2. (Gray (1936) presented his argument in terms of energy losses per unit *volume*, but it is simpler in terms of mass.) Gray then made the assumption—and it will be seen later (§8.4) that it is not quite true—that the ratio of the energy *absorbed* per unit mass of the medium to that of the gas was also s_{mg}. But the energy absorbed per unit mass of the gas is $J\ W_g/e$, where J is the charge per unit mass resulting from ionisation produced by the electrons and W_g is the average energy required to produce an ion pair in the gas. The absorbed dose, D_m, in the medium is thus given by

$$D_m = s_{mg}\ J\ W_g/e, \qquad (8.1)$$

and this is the Bragg–Gray relation in modern terminology.

Gray's cavity theory effectively requires or assumes:

(a) that charged particle equilibrium exists in the absence of the cavity;

(b) (i) that the cavity does not disturb the charged particle fluence or its distribution in energy and direction, and (ii) as a corollary to (i) that charged particle production in the cavity is negligible or no different from that in an equal mass of surrounding medium;

(c) that the mass stopping power ratio, s_{mg}, does not vary with energy; and

(d) that the secondary charged particles lose energy by a process of continuous slowing down, i.e. by a large number of very small energy loss events.

It is fortunate for radiation dosimetry that the mass stopping power ratio, s_{mg}, is usually only a slowly varying function of electron energy, so that the actual energy distribution does not need to be known with great accuracy. Improvements in cavity theory since 1936 have largely centred on more detailed considerations of what it is appropriate to use for s_{mg} in equation (8.1).

8.2 Early Improvements in Mass Stopping Power Ratio

Laurence (1937) allowed in his theory for the energy spectrum of electrons as they slowed down in the medium and for the need to average the stopping power ratio over this energy spectrum. Whyte (1954) modified Laurence's theory to accommodate the density effect, that is the effect on the stopping power caused by polarisation in the medium near the tracks of charged particles (§2.5.4). Others, for example Cormack and Johns (1954), pointed to the need to use only the 'collision' part of the stopping powers and to set aside the 'radiative' part, i.e. the energy lost by the electrons in bremsstrahlung production. Although this resulted in energy loss by the electrons it played no part in the energy absorbed by the medium or the gas in the vicinity of the radiative event.

8.3   Fano's Theorem

Gray had appreciated that his assumption that the presence of the cavity did not affect the electron fluence or its distribution in energy or direction was most likely to hold if the gas in the cavity had a similar composition to the surrounding medium. Many practical ionisation chambers had used this idea. Fano (1954) put this matter on a secure theoretical foundation by providing a formal mathematical proof of a theorem which stated 'In a medium of given composition exposed to a uniform flux of primary radiation (such as x-rays or neutrons) the flux of secondary radiation is also uniform and *independent* of the density of the medium as well as of the density variations from point to point.' In particular the introduction of a low-density cavity into a material would not alter the detailed electron fluence providing the material in the cavity has the same elemental composition, and *in these circumstances there is no limitation put on the size of the cavity*. This is extremely important. It means that providing we have a homogenous cavity, i.e. one having the same elemental composition as the surrounding material, we no longer have to worry about the secondary charged particles losing only a small fraction of their energy in crossing the cavity. With low-energy x-rays and even high-energy neutrons it is not usually possible to make a gas-filled cavity small enough to meet these conditions and still contain enough gas to give adequate sensitivity of measurement. Measurement of these radiations rests on the use of homogeneous ionisation chambers and the application of Fano's theorem.

This theorem does, however, require further comment. Although not specifically stated by Fano it has been pointed out by, for example, Roesch (1968) and Harder (1974) that the theorem applies to an infinite medium. It is not true near the interface between two different media where charged particle equilibrium has not been established. Harder (1974) has added appropriate boundary conditions.

Also the theorem assumes that the mass stopping power of a medium is independent of its density. Unfortunately, the so-called density effect—see Chapter 2—leads to deviations from such independence, but some allowance can be made for this (Whyte 1954). This remains an important theorem justifying the widespread use of homogeneous cavities.

8.4 Allowance for Large Energy Losses

An important advance took place in 1955. Up until then all theories had tacitly assumed that the electrons lost their energy in a very large number of very small energy loss events, that is that the electrons slowed down 'continuously'. Such an assumption meant that the energy lost by the electrons was absorbed 'on the spot' by the material through which the electrons were passing. Spencer and Attix (1955) and Burch (1955) both drew attention to the part played by δ rays, that is the electrons released by the high-energy electrons with sufficient energy to produce further ionisation on their own account. Some of these electrons released in the gas cavity would have sufficient energy to escape from the cavity carrying some of their energy with them. This would reduce the energy absorbed in the cavity and would require modification to the stopping power of the gas. Spencer and Attix (1955) suggested that the mass stopping power should be calculated using only collisions in which the energy loss was less than some cut-off value, Δ, where Δ was the energy an electron would have to have for its projected range to approximate to the dimensions of the cavity. Only energy losses less than Δ were considered to give rise to energy deposition within the cavity. Such mass stopping powers are called 'restricted' stopping powers and are written, $(S/\rho)_\Delta$. In this notation Δ is usually assumed to be expressed in eV. It was necessary to add the δ rays of energy greater than Δ to the spectrum of high-energy electrons (primaries and secondaries) in the calculation of the average stopping power ratio. Correspondingly, the lower limit of integration over the electron energy spectrum when calculating the stopping power ratio was set at Δ. In effect the secondary electrons were directed into two groups—those with energy less than Δ were considered to deposit all their energy in the cavity, and those with energy greater than Δ were considered to deposit no energy in the cavity and were added to the spectrum of electrons that were slowing down. It was further assumed that the electrons with energies less than Δ that crossed the cavity boundary in both directions did not result in a net energy transfer. This assumption is less likely to hold as the difference in the atomic numbers of the wall material and cavity increases. Spencer (1965) later modified the theory to include the density effect discussed in §2.5.4.

In contrast to previous theories, that of Spencer and Attix indicated that the ionisation per unit mass of gas would be a function of the mass of gas, i.e. of the volume of the cavity or the pressure of the gas in it. This function would also be expected to vary for different combinations of medium and gas. These predictions were open to experimental test and a number of investigators (Greening 1957, Attix *et al* 1958, Burlin 1961, 1962) found much better agreement with the Spencer–Attix theory than with the unmodified Bragg–Gray theory.

8.5 Photon Interactions in the Cavity

These cavity theories assume either that photon interactions in the cavity gas are negligible or that they are the same as would take place in the same mass of surrounding medium. Nahum (1978) has extended the Spencer–Attix theory to make approximate allowance for these photon interactions and he also reviews mass stopping power ratio calculations for the important materials water and air made to that date. Predictions from his theory have been given some experimental support (Nahum and Greening 1978), but have not as yet been tested for cavities with Δ greater than 500 keV.

We have already seen that when the cavity is very small and photon interactions within it are negligible the ratio of the energies absorbed per unit mass of the medium and of the gas is the same as an appropriately chosen mass stopping power ratio for electrons in the two materials. As the cavity increases in size photon interactions in it will become significant, and by the time the radius of the cavity exceeds the maximum range of electrons coming from the surrounding medium, the absorbed dose at the centre of the cavity will be determined entirely by these photon interactions and will be proportional to the mass energy absorption coefficient, $(\mu_{en}/\rho)_g$, of the cavity material. When the cavity is very large compared with the range of electrons coming from the surrounding material the average absorbed dose throughout the cavity will be determined by $(\mu_{en}/\rho)_g$ as the absorbed dose in the very thin 'rind' of the cavity adjacent to the surrounding medium will be negligible. The absorbed dose in the bulk of the surrounding medium will be determined by its mass energy absorption coefficients, $(\mu_{en}/\rho)_m$, and the ratio of the absorbed doses in the medium and the cavity will be given by the ratio of their mass energy absorption coefficients, $(\mu_{en}/\rho)_m/(\mu_{en}/\rho)_g$. Thus as the size of the cavity increases the ratio of the absorbed doses in the medium and cavity gradually changes from s_{mg} to $(\mu_{en}/\rho)_m/(\mu_{en}/\rho)_g$. All this presupposes that the photon field is constant throughout both media.

Burlin (1966) has put foward a general cavity theory that covers the above points and tries to embrace cavities of all sizes. He applies a weighting factor, d, to a term including s_{mg} and a weighting factor $(1 - d)$ to a term including $(\mu_{en}/\rho)_m/(\mu_{en}/\rho)_g$, d gradually changing from 1 to 0 as the cavity size increases. The weighting factor is related to electron attenuation in the cavity. Burlin (1968) has also provided an excellent review of cavity theory up to that date.

8.6   Cavity Theory and Neutrons

Although cavity theory was developed initially for the dosimetry of photons, it is in principle equally applicable to the dosimetry of neutrons. Indeed Gray (1944) himself used it for this purpose. The problem, which is further discussed in §9.5 and §9.5.1 is that the ranges of recoil particles are usually so small that the requirements of the theory can only be met by using homogeneous chambers, that is by employing Fano's theorem. Matters are further complicated

(a) by the wide variety of charged particles arising in neutron irradiation;

(b) by the fact that W values for these particles differ and vary with energy in different ways; and

(c) by the need to use complex mixtures of gases in the cavities.

8.7   The Dosimeter Probe

Figure 8.1 illustrates schematically the generalised situation when a radiation sensitive dosimetry material, g, contained in a wall, w, is introduced into a material, m, in which the absorbed dose is required. Several special cases can be identified and it is convenient to deal with them in increasing order of complexity.

8.7.1   Case 1. All materials are the same

This is the ideal case. There are no interface problems or cavity size problems and no disturbance of the radiation field. No cavity theory is needed as absorbed dose is derived directly. It can only be applied in the rare case in which the material m is itself a suitable dosimetric

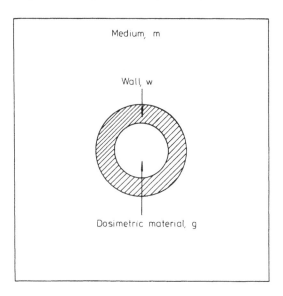

Figure 8.1　Illustrating a dosimetric probe.

material. Some very close approximations are possible, however. For example, if the absorbed dose is required in water it is possible to use the ferrous sulphate (Fricke) dosimeter (§11.3), which is 96% water and has an electron density 0.997 times that of water. If the wall is a plastic material such as polyethylene, which is $(H_2C)_n$ instead of H_2O, the deviations from complete uniformity are very small. (At low photon energies, where photoelectric absorption in the sulphur of the ferrous sulphate dosimeter becomes significant, the system deviates more from the ideal, but lower concentrations of sulphuric acid may be used.)

8.7.2 Case 2. All materials the same except for density

This is a special case of 1 above. We have a homogeneous probe inserted into material of the same elemental composition. Fano's theorem applies and the simple Bragg–Gray theory holds. An example of its application is found in neutron dosimetry when absorbed dose distributions in tissue are required. An ionisation chamber with a solid wall made of a tissue-equivalent plastic is filled with a tissue-equivalent gas and moved within a tank of tissue-equivalent liquid (see ICRU 1977).

8.7.3   Case 3. Wall, w, is the same as material, m

This reduces in effect to a cavity g in material m, and this situation has been discussed earlier in this chapter. The dosimeter response will be a function of cavity size. Providing the cavity g can be made small enough to be handled by the Spencer–Attix theory problems are slight. Consideration will need to be given to the extent to which the introduction of the foreign material g influences the distribution of radiation in m, but the fact that the cavity is small enough for the Spencer–Attix theory to be applied probably means that this disturbance will be small. If the size of g is too large for the normal Spencer–Attix theory to apply, it may be possible to use the extension by Nahum to allow for some interactions in the cavity or to use the general theory of Burlin.

8.7.4   Case 4. Wall, w, of same material as dosimetric material, g

If the thickness of the wall, w, exceeds the range of electrons ejected by uncharged particles from the medium, m, the dosimetric material, g, will respond as if it were surrounded solely by material g. Its response will then be conditioned by its mass energy absorption coefficient and the absorbed dose in the medium is obtained using the ratio of mass energy absorption coefficients of m and g. This is an important case widely employed in dosimetry. As in case 2 above, w and g need not be in the same phase, i.e. w can be a solid and g a gas or liquid, and we will still have what is called a homogeneous detector. A common example of this is the air-walled air-filled cavity for the measurement of exposure. Absorbed dose is derived from the exposure measurement as indicated in Chapter 7.

The use of a wall having the same composition as the dosimetric material is also of value in the dosimetry of electrons with initial energies in the range 5–50 MeV. The spectrum of electrons with energies below 0.1 MeV stays fairly constant with depth in the irradiated material, but depends on the nature of that material. There is therefore a change in this spectrum at the boundary between the medium, m, and the wall, w, and at the boundary between the wall and the dosimetric material, g. It is desirable that the spectrum entering g is characteristic of g so that the appropriate stopping power can be determined. A very small thickness of wall of the same material as g—say 2 mg cm^{-2}—is sufficient to bring this about. For constructional reasons the wall may need to be of greater thickness than this. As the spectrum changes in such a small thickness the problem is only important when 2 mg cm^{-2} is a significant fraction of the thickness of g. This is unlikely to be the case with solid or liquid detectors, but gas-filled ionisation chambers are a different matter. (See discussion in ICRU 1984a.)

If the wall thickness is less than the range of electrons ejected from the medium the position is complex. The dosimetric material will receive some electrons from the medium and some from the wall and some from within itself. Such arrangements are best avoided, but a great deal of the world's high-energy photon dosimetry has been carried out with this system and in consequence it has received attention from Nahum and Greening (1978). They deal in detail with the gradual change in the balance of the electrons reaching an air cavity from an air-equivalent wall and a water-equivalent surround medium as the photon energy increases from that corresponding to x-ray production at 2 MV to that at 31 MV (see also §9.2.4).

8.7.5 Case 5. Medium m, wall w, and dosimetric material, g, all different

The general case is even more complex than that considered in the previous paragraph. It has been examined by Almond and Svensson (1977) and Shiragai (1978) (see also §9.2.4) and their treatments are readily simplified to deal with the case in which w and g are the same. In the general case if the wall thickness exceeds the range of electrons released in the medium, and if g is sufficiently small for the Spencer–Attix theory to apply, we can use that theory to obtain the absorbed dose in the wall material. We can then calculate the absorbed dose in the medium m by using the ratio of the mass energy absorption coefficients in the case of photon irradiation or the mass stopping powers in the case of electron irradiation. Nevertheless there is added complexity in trying to allow for the disturbing effects of the many materials on the photon or electron fields.

8.8 Cavity Chamber Exposure Standards

In Chapter 5 it was indicated that exposure standards based on the free-air chamber principle were difficult to realise for x-rays generated above about 300 kV. Instead cavity chambers were used, and now that cavity theory has been discussed we can give consideration to exposure standards for use above 300 kV.

The basic Bragg–Gray relationship has already been given in equation (8.1) as

$$D_m = s_{mg} J W_g/e.$$

From equation (7.2) D_m can also be expressed as

$$D_m = X \frac{W_{\text{air}}}{e} \frac{(\mu_{\text{en}}/\rho)_m}{(\mu_{\text{en}}/\rho)_{\text{air}}}.$$

Equating these expressions for D_m, the exposure, X, is given by

$$X = J \, s_{\text{mg}} \frac{W_g}{W_{\text{air}}} \frac{(\mu_{\text{en}}/\rho)_{\text{air}}}{(\mu_{\text{en}}/\rho)_m}. \tag{8.2}$$

It should be emphasised that both the expressions for D_m require charged particle equilibrium to be established and therefore this condition also applies to equation (8.2). This equation shows that exposure can be measured with a cavity chamber if the terms on the right hand side can be measured or otherwise evaluated. If the cavity wall and the filling gas are air equation (8.2) reduces to

$$X = J, \tag{8.3}$$

and we have the great simplification of the free-air chamber. It is not possible to make a perfect 'solid air' wall, but we can easily use air as the filling gas. In this case

$$X = J \, s_{\text{mg}} \frac{(\mu_{\text{en}}/\rho)_{\text{air}}}{(\mu_{\text{en}}/\rho)_m}, \tag{8.4}$$

and there is no need to know W_{air} and W_g, data which are required in the general case of equation (8.2). If, furthermore, the wall material is chosen so that s_{mg} and $(\mu_{\text{en}}/\rho)_{\text{air}}/(\mu_{\text{en}}/\rho)_m$ are close to unity and accurately known, a cavity chamber standard of exposure can be realised. All standards laboratories appear to have chosen graphite as the material from which to make their cavity chambers.

These exposure standards are used to give the exposure at a point in the absence of the chamber. But in order to establish full electronic build-up (and a close approximation to electronic equilibrium) the chambers have a wall thickness of 3 mm or more for ^{60}Co radiation. The chamber wall will then both attenuate and scatter the primary beam and a correction of about 1.5% has to be made for this.

Because of the essentially forward ejection of the Compton recoil electrons giving rise to the measured ionisation in the cavity, the photon interactions producing these electrons will have taken place, on average, at a point within the front wall of the chamber. This means that the photon attenuation in the *full* thickness of the front wall does not have to be applied. It also gives rise to uncertainty as to the point to which the exposure measurement should be attributed as the electrons entering the air cavity will have arisen at different points within the front wall of the chamber.

As the photon energy increases above that of ^{60}Co radiation the ranges of electrons increase rapidly and this problem becomes more and more serious, so much so that the concept of exposure is not used above about 3 MeV. Further small corrections have to be applied to the measurement but for details of these and the construction of the cavity exposure standards reference should be made to the original publications. For the BIPM chamber see Boutillon and Niatel (1973); for the NBS chambers see Loftus and Weaver (1974); and for the NPL chamber see Barnard *et al* (1964).

9 Comparison of Electron, Photon and Neutron Dosimetry

9.1 The Common Ground

The dosimetry of electrons, photons and neutrons has to some extent been developed by different groups of people, and in consequence different aspects have been emphasised for different radiations. However, the same basic principles apply to all these radiations and the treatment which follows is intended to show the common ground before going on to discuss special cases and points of difference.

The great majority of precise measurements of absorbed dose, whether arising from electron, photon or neutron irradiation, are made using gas-filled ionisation chambers. For all measurements of electrons and neutrons, and for measurements of photons with energies higher than those of ^{60}Co γ rays, the assumption is made that the ionisation chambers obey the Bragg–Gray relationship or the extensions of it discussed in Chapter 8. We will first consider the general case of a chamber with a wall material, w, filled with a gas, g. The wall is assumed to be of sufficient thickness for approximate charged particle equilibrium to be established during photon calibration. The special cases of chambers for electron or photon or neutron radiation will be derived later.

From the Bragg–Gray equation the absorbed dose D_w in the wall is given by

$$D_w = J_g \left(\frac{W_g}{e}\right)_r (s_{wg})_r, \qquad (9.1)$$

where $J_g = Q/m$ the charge per unit mass in the gas, g, $(W_g/e)_r$ is the average energy per ion pair in the gas g for radiation r and $(s_{wg})_r$ is the mass stopping power ratio of the wall material and the gas for the charged particles arising from radiation r. The last two terms in equation (9.1) are usually known with reasonable accuracy, but J_g is unknown. It can be obtained by determining the charge sensitivity of the measuring system of the dosemeter, and by careful metrology of the chamber dimensions from which m can be calculated. Indeed this is what is done by standards laboratories for their cavity exposure standards (§8.8) and has been done at a standards laboratory for a commercial chamber (Henry 1979). Normally, however, J_g is obtained by exposing the chamber to a known exposure X of photons of sufficient energy for the chamber to act as a Bragg–Gray cavity. Suppose the dosemeter reading is M. Then the exposure calibration factor of the dosemeter is $X/M = N_c$, say, and $X = MN_c$.

As a result of this exposure the absorbed dose in the chamber wall $(D_w)_c$ is given by equation (7.2) (the subscript c is used to indicate calibration conditions)

$$(D_w)_c = MN_c \left(\frac{W_{air}}{e}\right)_c \left(\frac{(\mu_{en}/\rho)_w}{(\mu_{en}/\rho)_{air}}\right)_c . \tag{9.2}$$

From the Bragg–Gray relationship the absorbed dose in the gas, g, will be $(D_w)_c/(s_{wg})_c$ and will give rise to a charge per unit mass J_g given by

$$J_g = \frac{(D_w)_c}{(s_{wg})_c (W_g/e)_c} . \tag{9.3}$$

Using equation (9.2) in (9.3)

$$J_g = MN_c \left(\frac{W_{air}}{e}\right)_c \left(\frac{(\mu_{en}/\rho)_w}{(\mu_{en}/\rho)_{air}}\right)_c \frac{1}{(s_{wg})_c (W_g/e)_c} , \tag{9.4}$$

and the ionisation per unit mass of gas has been found in terms of known or measured quantities.

If the chamber is subsequently placed in a material, m, and exposed to some other radiation, r, the absorbed dose in the gas, D_g, is obtained by multiplying the ionisation per unit mass, J_g, by the energy per ion pair in the gas for the new radiation, $(W_g/e)_r$. Thus

$$D_g = J_g \left(\frac{W_g}{e}\right)_r . \tag{9.5}$$

The absorbed dose, D_m, in the material, m, at the position of the ionisation chamber, but in the absence of the ionisation chamber, will be $(s_{mg})_r$ times that in the gas, i.e.

$$D_m = D_g (s_{mg})_r = J_g \left(\frac{W_g}{e}\right)_r (s_{mg})_r . \tag{9.6}$$

(This will only be true if the introduction of the chamber into the medium does not alter the particle fluence. In general the fluence *will* be altered to some extent and corrections have to be made to the measurements to allow for this. See §9.2).

Using equation (9.4) in (9.6)

$$D_m = MN_c \left(\frac{W_{air}}{e}\right)_c \left(\frac{(\mu_{en}/\rho)_w}{(\mu_{en}/\rho)_{air}}\right)_c \frac{(s_{mg})_r}{(s_{wg})_c} \frac{(W_g/e)_r}{(W_g/e)_c}. \tag{9.7}$$

This is the general case. Some special cases are of interest. For the measurement of both photons and electrons it is the practice to use air as the cavity gas and most commonly the absorbed dose to water is required. Then equation (9.7) reduces to

$$D_{water} = MN_c \left(\frac{W_{air}}{e}\right)_r \left(\frac{(\mu_{en}/\rho)_w}{(\mu_{en}/\rho)_{air}}\right)_c \frac{(s_{water\ air})_r}{(s_{w\ air})_c}. \tag{9.8}$$

If the wall material is air-equivalent at the calibration quality

$$D_{water} = MN_c \left(\frac{W_{air}}{e}\right)_r (s_{water\ air})_r. \tag{9.9}$$

For neutrons it is usual to determine the absorbed dose in tissue. The chamber wall is made of conducting tissue-equivalent plastic and is filled with a tissue-equivalent gas. In this case equation (9.7) reduces to

$$D_{TE} = MN_c \left(\frac{W_{air}}{e}\right)_c \left(\frac{(\mu_{en}/\rho)_{TE}}{(\mu_{en}/\rho)_{air}}\right)_c \frac{(W_{TE}/e)_N}{(W_{TE}/e)_c}, \tag{9.10}$$

if it is assumed—as is usually done—that the mass stopping power ratio of TE plastic and TE gas is unity both at the calibration quality and for the recoil protons and heavier charged particles arising from the neutron irradiation. If the TE plastic does not quite correspond to the tissue of interest a small adjustment may be necessary.

The above presentation has been made in terms of exposure, but could equally well have used air kerma. If N_k is the air kerma calibration factor for the ionisation chamber obtained after being subjected to a known air kerma, then $N_k = N_c(W_{air}/e)_c [(\mu_{tr}/\rho)/(\mu_{en}/\rho)]_{air}$ (see, for example, equation (6.3)). Thus in equations (9.2), (9.4), (9.7), (9.8) and (9.10) substitute N_k for $N_c(W_{air}/e)_c$ and replace $(\mu_{en}/\rho)_{airc}$ by $(\mu_{tr}/\rho)_{airc}$.

9.1.1 C_E, C_λ, N_D **and** N_{gas}

The product of the terms by which MN_c in equations (9.8) and (9.9) has to be multiplied in order to give the absorbed dose in water has often been called C_λ, since first used by Greene (1962) for photons, or C_E following its use by Almond (1967) and Svensson and Petterson (1967) for electrons. In deriving values of C_E and C_λ some of the correction factors discussed in §9.2 below have always been included. These symbols were adopted by ICRU (1972, 1969) and values were recommended. However, in the calculation of C_λ it was, in effect, assumed that the ionisation chamber had a water-equivalent wall whereas in deriving C_E the chamber was assumed to have an air-equivalent wall. The application of these values to the *same* chamber when used to measure both photons and electrons led to inconsistent dosimetry as pointed out by Nahum and Greening (1976, 1978) and Almond and Svensson (1977).

The right hand side of equation (9.8) contains terms with subscripts 'water' and 'air'. These terms are common to all ionisation chambers. However, there are other terms with the subscript 'w'. This signifies that the term relates to a quantity which is a property of the wall material of the chamber. These terms are *not* common to all chambers. It must be emphasised that C_E and C_λ are specific to a particular type of chamber. This is so, not only for the reason given above, but also because the correction factors discussed in the next section, and which have to be included in C_E and C_λ, depend on the size, shape and material of the chamber.

Values of C_λ for the particular ionisation chamber used in the United Kingdom for the dissemination of the NPL standard have been published by the Hospital Physicists' Association (HPA 1983).

An alternative method of dealing with the basic equations in §9.1 is adopted in North America (AAPM 1983), in Scandinavia (NACP 1980, 1981) and by the ICRU (1984a). All three derive expressions for the absorbed dose in the gas of the ionsation chamber per unit reading of the calibrated exposure meter. This quantity is called N_{gas} by the AAPM and N_D by NACP and ICRU. From equation (9.4) we obtain J_g, the ionisation per unit mass of gas. If this is multiplied by W_g/e we have the absorbed dose in the gas, and if we put the meter reading $M = 1$, then

$$N_D = N_c \left(\frac{W_{air}}{e}\right)_c \left(\frac{(\mu_{en}/\rho)_w}{(\mu_{en}/\rho)_{air}}\right)_c \Big/ (s_{wg})_c.$$

Not only does N_D share with C_λ and C_E the property of being specific to a *type* of ionisation chamber, but it also has the property of being unique to an *individual* chamber, as it includes the exposure calibration factor. Thus, no values for N_D can usefully be tabulated.

Just as with C_λ and C_E, N_D must include the correction factors discussed in the next section.

9.2 Correction Factors

For accurate dosimetry several small corrections have to be applied to equations such as (9.7) and its derivatives. The literature of these corrections is confusing as many different symbols have been used for the same correction and sometimes the same symbol for different corrections. Furthermore words such as 'displacement' and 'perturbation factor' have been used by different authors in somewhat different senses. The main corrections are given below with an indication of their magnitudes for commonly used ionisation chambers. (See also AAPM 1983.)

9.2.1 Attenuation and scattering in the chamber wall during calibration in air against an exposure meter

The exposure meter will have been calibrated to indicate the exposure *in the absence of the dosemeter*. However, when a dosemeter is being calibrated to obtain J_g it is necessary to know the absorbed dose on the *inside* of the ionisation chamber. There will be attenuation of the primary radiation in the front wall of the chamber and this will be compensated to some extent by scattering of radiation from all walls into the gas cavity. The chamber wall has to be thick enough to establish approximate charged particle equilibrium and for ^{60}Co radiation this leads to a correction factor of typically 0.98–0.99 being required on the right-hand side of equation (9.7) and its derivatives (Barnard *et al* 1959, Johansson *et al* 1978, Almond *et al* 1978b, Nath and Schulz 1981).

9.2.2 Dissimilarity of chamber wall and phantom material

When the chamber of the dosemeter of known J_g is introduced into a water phantom (or some other material) the chamber wall and build-up cap may attenuate and scatter the radiation in a different manner from the water they displace. The effect can be minimised by removing the build-up cap leaving only the chamber wall which is typically about 0.5 mm thick. The resulting correction is likely to be very small. The lead to the chamber may also produce some disturbance of the radiation fluence, but this can be allowed for if necessary by observing the effect of introducing a dummy lead in addition to the real lead.

9.2.3 Effect of gas cavity

The introduction of a gas cavity into the water (or other) phantom will alter the radiation fluence because of the different attenuations and scatterings in the water and gas. In the case of photon measurements interest usually focuses on depths at which quasi electronic equilibrium has been established. In this region the absorbed dose falls approximately exponentially with depth, with the result that a given increase of depth produces a constant factor reduction in absorbed dose. The air cavity of the chamber therefore affects the measured dose by the same factor irrespective of its position in the medium. However, the amount by which the factor differs from unity might be expected to be proportional to the linear dimensions of the cavity and to the absorbed dose gradient. The latter point means that the correction factor will be a function of the radiation quality. The correction factor is less than unity by about 0.3%/mm of air cavity radius for ^{60}Co radiation falling to 0.2%/mm for 30 MV radiation (Cunningham and Sontag 1980). The centre of the chamber is taken to be the point of measurement.

The same approach is adopted for neutrons, but unfortunately appropriate data are very limited, the correction factor being less than unity by roughly 0.4%/mm of chamber radium (Williams *et al* 1982, Awschalom *et al* 1983). The use of chambers of small diameter will minimise uncertainties in this correction factor.

In electron dosimetry there is usually interest in the region where absorbed dose is increasing with depth as well as that where it is decreasing. A correction factor would vary with absorbed dose gradient and would change sign when the point of measurement passed from one of the above regions to the other. In consequence it is more appropriate to correct for the effects of the air cavity (at least in part) by considering the effective point of measurement to be shifted towards the source of radiation. For a flat coin-shaped chamber it is moved to the inner surface of the front wall of the chamber and for a thimble-shaped chamber the effective point of measurement is about 0.5 of the radius in front of the chamber centre. For further details see Johansson *et al* (1978).

With electrons there is a further effect which is absent with photons. The gas cavity does not scatter electrons as much as the surrounding denser material. In consequence the electron fluence in the cavity may be greater or smaller (depending on the cavity geometry) than that in the undisturbed medium (for theory see Harder 1968, Mandour and Harder 1977, Svensson and Brahme 1979, for experiment see Johansson *et al* 1978). The correction factor is likely to be more than 0.99 for thimble-shaped chambers with diameters less than 7 mm with electrons of energy greater than 15 MeV, but can fall to 0.95 for 2 MeV electrons. The effect does not occur with photons under conditions of electron equilibrium because the Fano theorem (§8.3) proves that the introduction of the low-density cavity does not disturb the secondary electron fluence.

9.2.4 Change of effective wall material with energy during photon irradiation

Equation (9.7) strictly gives the absorbed dose in the material m only if the chamber wall has not affected the measurement. This will only be true if the chamber wall is the same as the surrounding material m. At the calibration quality *all* electrons crossing the cavity in fact come from the chamber wall and its build-up cap (the chamber acts as a 'photon detector'), but as the photon energy increases so also do the secondary electron ranges and eventually most of the electrons crossing the cavity have arisen as the result of photon interactions with the surrounding medium (the chamber is acting as an 'electron detector'). Removing the build-up cap minimises the number of electrons arising in the chamber rather than the surrounding medium, but some effect remains. A brief account of the theory is given below.

Suppose we have an air-filled cavity and the chamber of wall material w is placed in a water phantom. The absorbed dose to water is required. We measure the ionisation per unit mass of air, J_{air}, and multiplying by W/e gives the energy per unit mass in the air cavity, D_{air}. The energy per unit mass of water, D_{water}, placed at the same point as the air cavity would be $(s_{water\ air})_\lambda$ times this, where λ indicates photon energy,

i.e.
$$J_{air}\,\frac{W}{e} = D_{air} = D_{water}\,(s_{air\ water})_\lambda, \tag{9.11}$$

(note that $(s_{air\ water})_\lambda = 1/(s_{water\ air})_\lambda$).

But if electrons crossing the cavity arise entirely from photon interactions with the wall material we have

$$J_{air}\,\frac{W}{e} = D_{air} = D_w\,(s_{air\ w})_\lambda = D_{water}\left(\frac{(\mu_{en}/\rho)_w}{(\mu_{en}/\rho)_{water}}\right)_\lambda (s_{air\ w})_\lambda. \tag{9.12}$$

In practice a fraction β of the ionisation will arise from electrons released in water and a fraction $(1 - \beta)$ from electrons released in the wall material. Thus

$$J_{\text{air}} = \beta J_{\text{air}} \text{ from water} + (1 - \beta) J_{\text{air}} \text{ from wall}. \quad (9.13)$$

Using equations (9.11) and (9.12)

$$J_{\text{air}} \frac{W}{e} = D_{\text{water}} \left[\beta (s_{\text{air water}})_\lambda + (1 - \beta) \left(\frac{(\mu_{\text{en}}/\rho)_{\text{w}}}{(\mu_{\text{en}}/\rho)_{\text{water}}} \right)_\lambda (s_{\text{air w}})_\lambda \right]. \quad (9.14)$$

If the wall material, w, is water equivalent the complicated expression in square brackets reduces to $(s_{\text{air water}})_\lambda$ and equation (9.14) reduces to equation (9.11). There is therefore substantial advantage in using a water-equivalent ionisation chamber for high-energy photon measurements when absorbed dose in water is required. If the wall material is air-equivalent, equation (9.14) reduces to

$$D_{\text{water}} = J_{\text{air}} \frac{W}{e} \left[\beta (s_{\text{air water}})_\lambda + (1 - \beta) \left(\frac{(\mu_{\text{en}}/\rho)_{\text{air}}}{(\mu_{\text{en}}/\rho)_{\text{water}}} \right)_\lambda \right]^{-1}. \quad (9.15)$$

This also is an important case as many ionisation chambers have been designed to be close to air equivalence. Nahum and Greening (1978) have discussed this problem and have given values of $(1 - \beta)$ for a chamber of wall thickness 74 mg cm^{-2} (after removal of build-up cap). Their values, expressed as a percentage, are shown in figure 9.1 and are based on cavity theory and experiment. Experimental results for various chambers showing the effect of the gradual transition of effective wall material with photon energy have been given by Johansson *et al* (1978), Almond *et al* (1978b) and Lempert *et al* (1983). The theory has been discussed by Shiragai (1978). The effect leads to corrections of up to 4% to equations (9.7) or (9.9) if an air-equivalent chamber is used in water. Effects could be greater with more dissimilar materials.

In principle similar problems could arise with neutrons, but in practice it has been usual to employ tissue-equivalent chambers and to use them in tissue-equivalent phantoms, or to have an adequate wall thickness to ensure that it was neutron interactions with tissue-equivalent material that produced the charged particles crossing the gas cavity. Problems

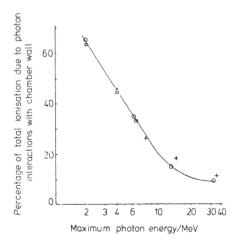

Figure 9.1   Variation with maximum photon energy of the percentage of the ionisation in a chamber with air-equivalent walls of thickness 74 mg cm^{-2} immersed in water, that is due to photon interactions with the walls. $\bigcirc$ from modified cavity theory with $\Delta = 0.5$ MeV; $\square$ from measurements without build-up cap; $+$ assuming build-up curve is exponential (from Nahum and Greening 1978).

could arise if the tissue-equivalent chamber wall had a thickness less than the range of the secondary charged particles and the chamber was placed in a material that was not tissue-equivalent.

9.2.5   Recombination and mass of gas

When ionisation chambers are calibrated at standards laboratories it is usual to use quite low exposure rates and steady sources of radiation. In these circumstances it is not difficult to collect virtually all the ions that are formed. When secondary standards are used to calibrate field instruments or when the field instruments are used subsequently, exposure rates may well be high and the radiation may come in intense pulses. In such conditions all the ions may not be collected and a correction must be made. The problem can be particularly severe in neutron dosimetry where heavy ions cause densely ionised tracks from which full extraction of charge is difficult. The correction for recombination—or lack of saturation—is briefly discussed in §11.2.1.

All measurements made with ionisation chambers that are not hermetically sealed need to be corrected to the temperature and pressure for which the calibration applies. The reading obtained experimentally is proportional to the mass of gas in the chamber which, in turn, is directly proportional to the pressure and inversely proportional to the absolute temperature.

9.3 Electron Dosimetry

In principle electrons present the least problem of the three particles discussed in this chapter. They are directly ionising and at low energies give rise only to other electrons. At higher energies some bremsstrahlung is produced but a 1 MeV electron loses less than 0.5% of its energy in this way in slowing to rest in water. A 10 MeV electron loses 4% and a 30 MeV electron 12% of its energy in this way. The bremsstrahlung will in its turn produce further electrons and some positrons. Photonuclear and electronuclear reactions are usually small. Mainly one type of charged particle is therefore involved in electron dosimetry.

When the electron beam passes from one medium to another the fast electron fluence and spectrum are unchanged—except of course for a continuing loss of energy—and a new low-energy spectrum ($E <$ 0.1 MeV) is very rapidly established. With photons and neutrons, on the other hand, the charged particle fluence and spectrum will change on change of medium due to differing interaction cross sections of the photons or neutrons and the change will be spread over distances equal to the ranges of the charged particles produced.

The change of electron energy with depth in the irradiated material is quite rapid—about 2 MeV cm^{-1} in unit-density low-Z material—and this can produce appreciable changes in the ratio of the electron stopping powers of the medium, in which the absorbed dose is required, and the dosimetric material. The density effect can also contribute to changes in this stopping power ratio above about 1 MeV when a gas is the dosimetric material.

The aspect of electron dosimetry that presents more difficulty than with photon or neutron dosimetry is charged particle equilibrium. It is only in exceptional circumstances, such as within material containing a uniform distribution of a β-ray emitter, that electronic equilibrium exists in electron dosimetry. The fast electrons lose energy as they pass through an irradiated material and there is no corresponding generation of new fast electrons as there is in photon beams, or of other charged particles as with neutron beams. To minimise this problem electron dosimeters are usually made thin in the direction of the electron beam. It is this lack of electronic equilibrium that leads to the perturbation of the electron fluence when inhomogeneities of density occur in an irradiated material (see §9.2.3).

Any dosimeter placed in an electron beam will have some of the electrons terminating their tracks within it. Ionisation dosimeters will have some electrons stopping in the collecting electrode and these will be added to or subtracted from the true ionisation. Such dosimeters have to be designed to minimise this effect. It will usually be less serious in irradiation with indirectly ionising particles as at least quasi charged particle equilibrium frequently exists (§6.4), and as many charged particles leave the dosimeter as enter it. Chambers designed for electron dosimetry are shown in ICRU (1972), Morris and Owen (1975) and Mattsson *et al* (1981). For electron dosimetry protocols see NACP (1980) and AAPM (1983). For a very detailed discussion of electron dosimetry see ICRU (1984a).

9.4   Photon Dosimetry

Photons are indirectly ionising, that is they bring about their ionisation by a two-stage process. In the first stage, they interact with matter to produce electrons (and positrons) and these charged particles then produce ionisation along their tracks. The energy transferred from the photon beam to the irradiated material depends first on the photon interaction coefficients of the material. These coefficients vary rapidly with photon energy and, for most interactions, with the atomic number of the material. It is therefore important to choose dosimetric and/or 'wall' materials that have known ratios of interaction coefficients to those of the material in which the absorbed dose is required. If these ratios are not to be sensitive to the distribution of photon energies in the radiation being measured, the dosimetric material should have an atomic number close to that of the material of interest. Fortunately the photon interaction coefficients are a fairly smooth function of atomic number and photon energy and it is possible to produce mixtures of materials that match the interaction coefficients of another material but do not have the same elementary composition by weight. (See White 1977.) Once the electrons are produced by these interactions the problem reduces to that of electron dosimetry discussed above, except that there are some simplifications. A close approach to electronic equilibrium often exists, so there are not the same perturbation problems as with electrons. The energy of photons usually varies slowly with depth of penetration in the irradiated material so the spectrum of secondary electrons to which the photons give rise will also vary slowly. Stopping power ratios are not then so sensitive to position within the irradiated material as they are with electron irradiation.

When a photon beam passes through an interface between two materials there is a sudden change of electron production and a gradual change of electron fluence spread over a distance governed by the range of the secondary electrons. This problem is more serious than with electrons. For photon dosimetry protocols see NACP (1980), HPA (1983) and AAPM (1983).

9.4.1 Use of dosemeter calibrated in exposure

It was pointed out in §9.1 that photons with energies above those of ^{60}Co had to be measured by methods fundamentally the same as those adopted for electrons and neutrons. However, for photons with energies between about 5 keV and 1.25 MeV standards laboratories have exposure standards—free-air chambers or cavity exposure standards—against which dosemeters can be calibrated. In this respect the position for photons is different from that for electrons and neutrons. The derivation of absorbed dose via an exposure measurement has been discussed in Chapter 7. When the ionisation chamber of a calibrated exposure meter is placed in a water phantom (or other material) it should retain any build-up cap used during calibration and then the meter will indicate the exposure *in the absence of the ionisation chamber*. A correction is therefore required for attenuation and scatter in the water displaced by the chamber and cap. The correction depends only on the external dimensions of the chamber and cap and not on their internal construction. The correction is greater than that needed for the air cavity alone, discussed in §9.2 when the chamber was used as a Bragg–Gray cavity, but only because the outer dimensions are greater than the inner dimensions. For ^{60}Co radiation the correction factor is less than unity by about 0.3–0.4% per mm of external radius of the chamber. For a series of papers discussing this correction see Grant *et al* (1977), Almond *et al* (1978a), Cunningham and Holt (1978), Jayaraman *et al* (1979) and Holt *et al* (1979). For somewhat lower energy radiation the correction is smaller (i) because the build-up cap can be reduced or dispensed with entirely, and (ii) because Compton scattered photons retain a greater proportion of the initial photon energy and lead to greater compensation of primary radiation attenuation by scatter. (At very low energies where the photoelectric effect predominates this argument is no longer true.)

9.4.2 Low-energy photons

The discussions in §§9.1 and 9.2 have concentrated on photons produced by electrons accelerated to over 1 MeV or photons emitted by ^{60}Co. For such radiations many ionisation chambers can act as Bragg–Gray cavities. However, as the photon energy falls the ranges of the secondary electrons that are produced by the photons decrease rapidly, and Bragg–Gray conditions are not met. It is then necessary to utilise homogeneous chambers, i.e. chambers to which Fano's theorem (§8.3) can apply. Thus, if air is the filling gas the wall material must be equivalent to air in its radiation interactions.

At low photon energies the predominant interaction process is the photoelectric effect and it varies very rapidly with both photon energy and atomic number (see Chapter 2). Thus radiation detectors made of materials that are not closely matched to air will show a response per unit exposure or per unit air kerma that varies appreciably with photon energy. This energy dependence may lead to a need for greater knowledge of the spectral distribution in energy of the photons being measured, or may even require allowance to be made for the change in this distribution that can take place with changing depth in an irradiated material.

At low photon energies (say below 100 keV) bremsstrahlung production is negligible, so that the mass energy absorption coefficient and the mass energy transfer coefficient are equal. Furthermore, photon mean free paths are orders of magnitude greater than the ranges of the electrons to which they give rise, effectively establishing charged particle equilibrium even in the absence of photon equilibrium. In the light of these two factors (no bremsstrahlung but charged particle equilibrium) we can conclude that the kerma at a point in an irradiated medium is equal to the absorbed dose at the point. This is in contrast to the position with photons from a 6 MeV accelerator illustrated in figure 6.1.

For further details of the dosimetry of low-energy photons see ICRU (1970b) and Greening (1972).

9.5   Neutron Dosimetry

Neutrons like photons are indirectly ionising. The fluence of ionising particles giving rise to an absorbed dose is therefore dependent not only on the neutron beam but also on the probability of its interaction with matter. It was seen in Chapter 2 that the interactions of neutrons with matter show resonance peaks in elastic scattering and sudden changes in other interaction processes. Interaction probabilities are far more atom specific with neutrons than with photons and there are not such smooth variations of interaction coefficients with neutron energy and material atom number as there are with photons. There is, therefore, a necessity in neutron work for dosimetric materials to have the same atomic composition as the materials in which absorbed doses are required. It is particularly important that the hydrogen content is the same. Although the hydrogen content of soft tissue is only about 10% by weight, below 7 MeV the hydrogen contributes over 90% of the absorbed dose, and even at 18 MeV it produces about 70% of the absorbed dose.

In any material with even a small hydrogen content the ionising particles arising from neutron irradiation will be largely recoil protons. These have much smaller ranges than electrons of the same energy. In fact a 1 MeV neutron will give rise to a recoil proton with a mean range in tissue of only 0.01 mm, and even a 10 MeV neutron will only increase this range to 1.4 mm. As these ranges are so much less than those of the neutrons producing them, there is very close approximation to charged particle equilibrium. The situation is analogous to the dosimetry of low-energy x-rays. Because of the short ranges of the recoil protons and the even shorter ranges of the heavier recoil nuclei, it is not practicable to make heterogeneous ionisation chambers with cavities small enough to satisfy the Bragg–Gray requirements. It is necessary to resort to homogeneous chambers in which the chamber wall and the filling gas have the same atomic composition. In consequence the most commonly used ionisation chambers are made of conducting plastic with an atomic composition similar to that of soft tissue (in practice, most of the oxygen of tissue has to be substituted by carbon in the plastic) and the chamber is filled with a tissue-equivalent gas (again some compromises are necessary). Pioneer work in the production of these tissue-equivalent materials was done by Rossi and Failla (1956) and Shonka *et al* (1958), and the composition of the widely used plastic called A-150 has been discussed by Smathers *et al* (1977).

Fortunately there are two small simplifications in neutron dosimetry. As protons or heavier particles, rather than electrons, are the ionising particles there is negligible bremsstrahlung production, and the polar-isation (density) effect (§2.5.4) is of no consequence even with 1000 MeV protons and therefore has no influence on the stopping power ratios that have to be used.

9.5.1 *W* values

When electrons or photons are measured by ionisation methods the ionising particles are electrons and for these *W*, the average energy required to produce an ion pair in the gas, appears to be constant over the energy range of normal interest in dosimetry. But when neutrons are being measured the ionising particles are protons, α particles and heavy recoil nuclei and the gases used in the ionisation chambers can be mixtures more complex than the air usually used with electrons and photons. For these heavy charged particles and complex gases *W* is, in general, no longer constant as the particle energy varies, the values of *W* are not known with the accuracy that attaches to W_{air} for electrons, and *W* values in the chamber gas will differ for the heavy ionising particles and the electrons arising from the photon beams used in the chamber calibration. Furthermore the relative abundance of the heavy charged particles may change with neutron energy, bringing about a change in the value of *W* appropriate to a particular measurement.

From equation (9.10) it will be seen that in order to determine the absorbed dose in tissue-equivalent plastic it is necessary to know the ratio of W_N in tissue-equivalent gas, for the charged particles released by the neutrons, to the W_e in this gas for the electrons released by the photons used at calibration. The value of W_N in TE gas varies with energy (Goodman 1978, Goodman and Coyne 1980) but for neutrons between 1 and 20 MeV it is about 31.3 eV. Taking W_e in TE gas as 29.2 eV (Leonard and Boring 1973) gives W_N/W_e for this energy region as approximately 1.07. *W* for protons, α particles and some heavy recoil nuclei in other gases that may be used for neutron dosimetry can be derived from data in ICRU (1979a).

9.6 Mixed Field Dosimetry

Neutron dosimetry presents a difficulty not encountered in electron or photon dosimetry. Neutrons are always accompanied by photons once they enter the irradiated material. Even in the rather unlikely event of the primary neutron beam being free of photons, as soon as the neutrons start to interact with matter photons are produced by inelastic collisions (see §2.4.2) and as the neutrons are thermalised there are capture reactions, such as the important reaction with any hydrogen present, that give rise to photons. The consequence of this is that a problem of mixed field dosimetry is created. This dosimetric difficulty is resolved by making measurements with two dosimeters. The ideal would be to use one with a high sensitivity to neutrons and a low sensitivity to photons and another with a high sensitivity to photons and a low sensitivity to neutrons. In practice one dosimeter, T, is designed to have about the same sensitivity to photons and neutrons and the other, U, to have a lower sensitivity to neutrons than to photons. Measurements are made with both dosimeters in the same mixed neutron/photon field. Let R_T be the response (e.g. meter reading, scaler count, etc) of the dosimeter T, and t_N and t_G be its sensitivities (response per unit absorbed dose) to the neutrons and γ photons in the mixed field. Then

$$R_T = t_N D_N + t_G D_G. \tag{9.16}$$

Correspondingly, for dosimeter U, we will have

$$R_U = u_N D_N + u_G D_G. \tag{9.17}$$

In practice both dosimeters will have been calibrated in a known γ-ray field, e.g. ^{60}Co. Divide equation (9.16) by $(t_G)_{cal}$, the sensitivity of chamber T to this calibration γ ray, and equation (9.17) by the corresponding $(u_G)_{cal}$, then

$$\frac{R_T}{(t_G)_{cal}} = \frac{t_N}{(t_G)_{cal}} D_N + \frac{t_G}{(t_G)_{cal}} D_G, \tag{9.18}$$

and

$$\frac{R_U}{(u_G)_{cal}} = \frac{u_N}{(u_G)_{cal}} D_N + \frac{u_G}{(u_G)_{cal}} D_G. \tag{9.19}$$

Rewrite as

$$R'_T = k_T D_N + h_T D_G \tag{9.20}$$

and

$$R'_U = k_U D_N + h_U D_G, \tag{9.21}$$

where the meaning of the new symbols can be seen by comparison of the equations. Simultaneous solution of equations (9.20) and (9.21) gives

$$D_N = \frac{h_U R'_T - h_T R'_U}{h_U k_T - h_T k_U}, \tag{9.22}$$

$$D_G = \frac{k_T R_U' - k_U R_T'}{h_U k_T - h_T k_U}. \tag{9.23}$$

It is clear that many sensitivities are required. In practice some of these sensitivities are nearly equal and h_T, h_U and k_T are close to unity. There are advantages in making k_U very small. To achieve this, current practice in fast neutron dosimetry for radiotherapy and radiobiology employs a tissue-equivalent ion chamber filled with a tissue-equivalent gas as dosimeter, T, and a small geiger counter or non-hydrogenous ionisation chamber as dosimeter, U. Error analysis of equations (9.22) and (9.23) shows that it is difficult to get an accurate measurement of a small absorbed dose of photons accompanying the main neutron absorbed dose. Fortunately this is not usually necessary as the photons produce smaller biological effects than do neutrons for a given absorbed dose (for example, by about a factor of 3 for 7 MeV neutrons used for radiotherapy) and the effective error for dosimetric purposes is reduced accordingly. For a full discussion of mixed field dosimetry, calibration procedures, suitable dosimeters, and a wealth of other data relevant to neutron dosimetry, the reader is referred to ICRU (1977).

Protocols for neutron dosimetry have been published for Europe (Broerse *et al* 1981) and for North America (AAPM 1980). There are some differences in the recommended ionisation chambers, phantom materials and basic physical data (Mijnheer *et al* 1981) that will need to be resolved before comparable doses are recorded on both sides of the Atlantic. Ionisation chambers for neutron dosimetry have been the topic of a conference (Broerse 1980) and neutron dosimetry for radiotherapy has been reviewed by Greening (1983).

10 The Dosimetry of Radionuclides

10.1 The Quantity 'Activity' and Its Units

Before discussing the dosimetry of radioactive materials it is necessary to introduce the quantities and units in terms of which such materials are specified. These quantities and units have evolved over the years from somewhat vague concepts of very restricted application to a precisely defined quantity of general application with clear units in which it may be expressed. In 1910 a unit 'curie' was defined as 'the amount of radon in equilibrium with one gramme of radium', but what was meant by 'amount' was unclear. Later, in 1930, the application of the curie was widened to include any member of the uranium series of naturally radioactive elements all of which could, in principle, be in equilibrium with a mass of radium. The number of nuclear transformations per second which this 'amount' of material underwent was subject to experimental determination and fluctuated with time around the value 3.7×10^{10}. After the second world war the International Union of Physics and Chemistry redefined the quantity and the unit to embrace all radioactive materials and the definitions were expressed by the ICRU (1951) as follows: 'The curie is a unit of radioactivity defined as the quantity of any radioactive nuclide in which the number of disintegrations per second is 3.700×10^{10}.' Even at this time it was noted that a point of revision that could arise would be the number of curies in 1 g of radium. The ICRU (1962) introduced the new quantity *activity* saying, 'the activity (A) of a quantity of a radioactive nuclide is the quotient of ΔN by Δt where ΔN is the number of nuclear transformations which occur in this quantity in time Δt,' and where Δ had the meaning discussed in the section on the history of absorbed dose in Chapter 4. The special unit of activity was the curie defined as 3.7×10^{10} s^{-1} (exactly). In 1971 the ICRU dropped the Δs and used the differentials dN/dt, at the same time as it dropped the Δs from other quantities such as absorbed dose. The curie, which had until that time had the symbol c, was now represented by Ci in conformity with the convention that symbols for units derived from people's names should have a capital letter. The latest definition (ICRU 1980) has been further refined and reads, 'The activity, A, of an amount of a radioactive nuclide in a particular energy state at a given time is the quotient of dN by dt, where dN is the expectation value of the number of spontaneous nuclear transitions from that energy state in the time interval dt.' The 'particular energy state' is the ground state of the nuclide unless otherwise specified.

During the 1970s the ICRU canvassed world opinion on the introduction of SI units for certain radiation quantities, including activity. In SI the unit of activity is s^{-1} and as there would be very substantial difficulties in using such a unit, particularly when discussing rates of change of activity, the ICRU requested, and the General Conference on Weights and Measures sanctioned, a special name becquerel, Bq, for the unit s^{-1} when used as the unit of activity. The special unit of activity, the curie, remains as 3.7×10^{10} s^{-1} (exactly) but is to be phased out by about 1985.

10.2   The Air Kerma-rate Constant, Γ_δ, and Its Precursors

A group of further quantities needs to be discussed before the dosimetry of radionuclides is considered. In the early days of radium measurement Eve (1906) introduced a quantity, k, which for a time was known as Eve's number, and which related the production of ion pairs in air per unit time and volume, B, to the mass, m, of a point source of radium placed a distance l from the point of measurement in a very large volume of air. If attenuation and scatter in the air are ignored the inverse square law will hold and we have

$$B = km/l^2. \qquad (10.1)$$

In the units of that time it was usual to express k in ions/cm^3 g s at 1 cm. With the introduction of the roentgen for x-ray measurement in 1926 and the extension of its use to γ rays after 1937 (see Chapter 5) it was possible to express ions/cm^3 in roentgens and, with additional changes in the units of mass and time, to give k in units of R mg^{-1} h^{-1} at 1 cm. Furthermore, characterising the radium source by its activity rather than its mass, k could be expressed in units of R/mCi h at 1 cm.

With the extension of the curie for use with *any* radioactive nuclide, the ICRU (1951) suggested that all γ-ray emission be expressed in these same terms, namely R/mCi h at 1 cm. At that time no name was suggested for the quantity (although both then and since it has been widely referred to as the '*k* factor'), but the ICRU (1957) proposed that it be called the 'specific γ-ray emission.' Later (ICRU 1962) the name was changed to 'specific γ-ray constant' to focus attention on the *constancy* of the quantity for a particular radionuclide, rather than on the emission from a source. In fact the quantity did not turn out to be all that constant as the ICRU (1971a) changed both the quantity and the name. A change was made for the following reasons. Until that time reference had been made only to the γ-ray emission from radionuclides. But γ-ray sources may also emit characteristic x-rays and internal bremsstrahlung. The definition of the quantity was changed to include these additional photons (or at least those above a cut-off energy δ, which would be chosen according to the intended application) and the quantity was termed the exposure rate constant, Γ_δ, given by

$$\Gamma_\delta = \frac{l^2}{A}\left(\frac{\mathrm{d}X}{\mathrm{d}t}\right)_\delta, \tag{10.2}$$

where $(\mathrm{d}X/\mathrm{d}t)_\delta$ is the exposure rate due to photons of energy greater than δ (by convention expressed in keV) at a distance l from a point source of activity A.

One further change has been made to the quantity and its name by the ICRU (1980). With the adoption of SI in radiation dosimetry the quantity exposure is to be expressed in C kg^{-1}. But the exposure rate constant is most likely to be used to determine the absorbed dose in some material irradiated by a particular radionuclide, and the absorbed dose will be expressed in Gy. It is therefore convenient to substitute air kerma for exposure in equation (10.2) as kerma is expressed in Gy. Reference to equation (6.4) shows that air kerma, K_{air}, is obtained from exposure, X, by multiplying by $W_{air}/e(1-g)$, and as g, the fraction of electron energy lost as bremsstrahlung, will be only a fraction of 1% for the vast majority of radionuclides, it reduces substantially to multiplying exposure by a sensibly constant W_{air}/e. The new quantity is called the air kerma-rate constant, the symbol for which is still Γ_δ, but is now defined as

$$\Gamma_\delta = \frac{l^2}{A}\left(\frac{\mathrm{d}K_{air}}{\mathrm{d}t}\right)_\delta. \tag{10.3}$$

The SI unit for Γ_δ is m^2 J kg^{-1} which, when the special names gray and becquerel are used, becomes m^2 Gy Bq^{-1} s^{-1}.† Until the special units rad and curie are phased out it is possible to express the air kerma-rate constant in units of m^2 rad Ci^{-1} $s^{-1} = (10^{-12}/3.7)m^2$ J kg^{-1} (exactly).

A number of warnings must be sounded concerning the use of this quantity. It is defined in terms of an ideal point source. Any practical source will have finite size and will give rise to attenuation and scattering. Also annihilation radiation and external bremsstrahlung may be produced. Again, any medium between the source and the point of measurement will cause attenuation and scattering which will need to be corrected for. Many of these points have been discussed by Wyard (1955) and Dyson (1973).

10.3   External Sources

If the radioactive source or sources are external to the human body, or other organism or material in which absorbed dose is required, the dosimetric methods discussed in earlier chapters can normally be applied. Thus the dosimetry associated with a ^{60}Co source of very high activity (a few kCi or 100 TBq) is no different in principle from that associated with a linear accelerator producing high-energy photons. Furthermore, high-activity β-ray sources (Haybittle 1960), although not common in clinical practice, can be measured by methods similar to those used with accelerator-produced high-energy electrons.

A different approach is used with sealed sources such as radium needles, or modern substitutes, which are distributed in a geometrical arrangement over a surface which is parallel to and usually about 1–2 cm from the body surface. Here, although measurements could be made with small volume dosimeters, the dosimetry is usually done by calculation in a manner similar to that described in §10.4.3 where internal sealed sources are discussed.

10.4   Internal Sources

10.4.1   Dosimeter Probe

Occasionally when sealed or unsealed radioactive sources are internal to the body it is possible to make a direct assessment of absorbed dose

† The product of Bq and s = 1. This explains why the SI unit of Γ_δ, when special names are not used, appears to contain no unit of activity and no unit of time.

by inserting a dosimetric probe into the tissue of interest or by placing it against the surface of the tissue. An example is the insertion of a dosimeter into the rectum in order to measure the dose to the rectal mucosa during treatment of cervical cancer with sealed sources in the vagina and uterus. The technique is normally restricted to those cases where access to the tissue of interest can be obtained via a body cavity.

10.4.2 Empiricism

Although direct measurement is occasionally possible a much more common approach, particularly with unsealed sources, is frank empiricism. To have to make such a statement in a book on dosimetry is a confession of failure. The fact remains that the treatment of hyperthyroidism with radioactive iodine isotopes and the treatment of polycythaemia vera with ^{32}P have both been developed with little true knowledge of the absorbed doses delivered to the tissues of interest. In practice the *activity* of the material administered to the patient has been measured. This activity has been correlated with the clinically observed response of the patient, and has then been adjusted to bring about the required response in future patients. Unfortunately this procedure does not ensure that a particular absorbed dose will be delivered. For example, in the case of thyroid treatment individual patients will not take up the same proportion of the administered activity into the thyroid, the time course of the activity in the thyroid will not be the same, and the mass of the thyroid will vary from patient to patient. Problems of this kind pervade the whole subject of the dosimetry of unsealed internal sources of radiation, and it is for these reasons that the dosimetry of such sources, which is discussed in §10.4.4 below, has been almost entirely restricted to considerations of radiation hazards, and has had little impact on radiotherapy.

Even with sealed sources within the body where, in principle, the absorbed dose at any point can be determined with the accuracy required for radiotherapy (see §10.4.3), a measure of empiricism has existed. The reason for this is that the absorbed dose usually varies rapidly from point to point, and the choice of the particular point or points at which to assess the absorbed dose is a matter of judgment. However, this is no criticism of the dosimetry as such.

10.4.3 Calculation: Sealed Sources

For many years the most important sealed sources were radium needles, that is hollow tubes usually of platinum–iridium alloy, into which thin walled tubes containing radium salt were loaded and sealed. The early dosimetry of these sources owed much to the work of the Swedish physicist Sievert after whom the SI unit of dose equivalent is now named (see Chapter 12). He introduced a concept which he called the 'intensity-millicurie', ImCi, (Imc in Sievert's time) which was some kind of measure of the radiation field at 1 cm from 1 mCi (i.e. 1 mg) of radium filtered by 0.5 mm of platinum. By the use of appropriate conversion factors this concept can be expressed in terms of more modern quantities and units. Sievert assumed the inverse square law to hold, i.e. no attenuation or scattering of radiation, and then calculated the distribution of intensity-millicuries around various geometrical arrangements of radium sources (Sievert 1921, 1923, 1930, 1932). Figure 10.1 illustrates Sievert's method for a uniform line source of radium with the platinum filtration assumed to be a constant 0.5 mm Pt. Let the linear density of radium be ρ mg cm^{-1}. Then the ImCi at P, I_p say, is given by

$$I_p = \int_0^a \frac{\rho \, dx}{r^2}, \tag{10.4}$$

and as $dx = r \, d\theta/\cos \theta$ and $r = h/\cos \theta$

$$I_p = \rho \int_{\theta_1}^{\theta_2} \frac{d\theta}{h} = \frac{\rho}{h} (\theta_2 - \theta_1). \tag{10.5}$$

In practice, the filtration will vary with the angle of emission of the γ rays from the needle. In this case the radiation reaching P from the element dx will have been attenuated by $\exp(-\mu d/\cos \theta)$ where μ is the

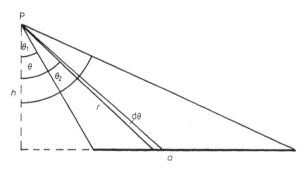

Figure 10.1   Calculation of radiation field round a linear radioactive source (filtration is assumed to be constant).

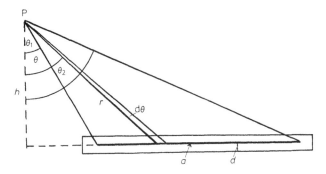

Figure 10.2 Calculation of radiation field round a linear radioactive source (allowance is made for oblique filtration).

linear attenuation coefficient of the needle wall and d is its radial thickness (see figure 10.2). Therefore

$$I_p = \frac{\rho}{h} \int_{\theta_1}^{\theta_2} \exp(-\mu d/\cos \theta)d\theta, \qquad (10.6)$$

and tables of these so-called Sievert integrals have been published by Sievert (1930, 1932). The region behind the ends of the needle needs different treatment but is usually of very little dosimetric interest.

Other workers at that time used similar methods to determine the radiation distribution around sources of various geometries, e.g. ring, disc and cylindrical sources. In particular, Parker, working in Manchester, investigated how best to distribute radium sources over a plane in order to produce uniform radiation fields in planes parallel to the source plane. His data were combined with the clinical judgment of Paterson and distilled into a few simple rules for distributing radium sources in a plane so that the radiation in adjacent planes of clinical interest would show no more than $\pm$ 10% variation from the mean. Other data were provided for calculating the actual 'doses' in these planes. The rules were applied initially to sources external to the body (Paterson and Parker 1934) but were subsequently extended to interstitial sources (Paterson and Parker 1938). Other people added to the system and much of this work, together with edited versions of original papers, has been very usefully assembled by Meredith (1967).

Although the advent of high-energy accelerators has reduced radiotherapy with surface applicators or interstitial sources, the above system is still in widespread use. With minor adjustments to allow for different specific γ-ray constants or k factors it can be used with sources other than radium (see Meredith 1967).

It was said earlier that Sievert's ImCi could be converted to other quantities and units by the use of appropriate factors. An important conversion was made by Parker in his dosimetry system. By assuming that the specific γ-ray constant or k factor for radium was 8.4 R/mCi h at 1 cm, he converted ImCi to exposure rate expressed in R h^{-1} and brought the dosimetry of radium sources into line with that of the x-ray sources of the time, for which 'dose' rates were expressed in R/unit time. Those who investigate the Paterson–Parker system further will discover that it specifies the number of mg h or mCi h needed for an exposure of 1000 R. It should be pointed out that a number of mCi h by itself says *nothing* about the exposure (or absorbed dose) of any tissue of interest. It is necessary also to know the spatial distribution of the sources relative to the point at which the exposure (or absorbed dose) is required. These further data are, in effect, incorporated in the rules, tables and graphs of the Paterson–Parker system.

If it is desired to bring the Paterson–Parker system into line with current dosimetry of other radiation sources, that is to state the treatment in terms of absorbed dose to water, then a number of conversion and correction factors must be applied:

(a) multiply R calculated by the Paterson and Parker system by 0.873 $[(\mu_{en}/\rho)_{water}/(\mu_{en}/\rho)_{air}]_{Ra}$ to obtain rad in water, or by 100 times less to get Gy in water (see equation (7.3));

(b) multiply R of the Paterson–Parker system by modern value of specific γ-ray constant (in R/mCi h at 1 cm)/8.4 (8.4 R/mCi h at 1 cm was the value used by Parker);

(c) correct for attenuation and scatter in the intervening medium (ignored by Parker); and

(d) allow for oblique filtration in wall of needle (not fully accounted for by Parker).

From figure 7.1 the ratio of the mass energy absorption coefficients of water and air is seen to be 1.112 over the energy range of the photons emitted by a sealed radium source (0.24 to 2.2 MeV). Thus the factor required for (a) above is $0.873 \times 1.112 = 0.970$. A modern value for the specific γ-ray constant of radium filtered by 0.5 mm Pt is 8.25 R/mCi h at 1 cm giving under (b) above a factor of 0.982. Attenuation of radium γ rays in tissue is about 1% per cm, and the correction required under

(d) is a reduction of 2 to 4%. The overall factor to convert Paterson–Parker R to rad in water is close to 0.90. For a detailed discussion see Shalek and Stoval (1968, 1969) and Stoval and Shalek (1968). The approach adopted by the 'Manchester School' has been described by Gibb and Massey (1980).

The availability of modern computers has very much reduced the labour involved in making the type of calculation undertaken by Sievert and Parker, but the principles are unchanged. Several radiotherapy treatment planning computer systems contain programs for calculating the absorbed dose at selected points in the vicinity of radium or other radionuclide implants when fed with data on the position of the radiation sources derived from radiographs of the implant taken from two positions. Such calculations are valuable in checking that the dose distribution planned in advance of the insertion of the radioactive sources, has in fact been achieved (Stoval and Shalek 1968).

With radium sources the construction is usually well established and the activity has been carefully compared with that of a standard. The specific γ ray constant or exposure rate constant has been extensively measured and great experience has been built up in the use of such sources. With other radioactive sources these considerations do not necessarily apply, and the uncertainties associated with the exposure rate constant (see end of §10.2) become relevant. It is better to use sources that have been specified in terms of the measured exposure rate at a specified distance (NCRP 1974, Dutreix and Wambersie 1975), or perhaps, nowadays, in terms of the air kerma rate at a specified distance. For a further discussion of these matters see also Jayaraman et al (1983).

10.4.4 Calculation: Unsealed Sources

The trouble with unsealed sources is that once they enter the body by injection, ingestion or inhalation, their whereabouts are not known in sufficient detail for accurate dosimetry. Thus any dosimetric calculations, however refined, are likely to founder through lack of adequate biological data. In consequence the great majority of such calculations are performed for radiation protection purposes, either to assess the likely hazards to people working with radioactive materials, to derive limits for concentrations of activity in air or water that people might breathe or drink, or to form an estimate of the risk to patients who may receive small quantities of radionuclides for diagnostic purposes.

A pioneering paper was published by Marinelli (1942) and was subsequently corrected and extended (Marinelli *et al* 1948). Further pioneering work was done by Mayneord (1950). Marinelli *et al* divided radiations into two categories, non-penetrating and penetrating. Non-penetrating radiation imparts all but a negligible fraction of its energy to the volume in which it is contained, and will comprise β rays, electrons and low-energy photons. Penetrating radiation, by contrast, deposits a significant proportion of its energy outside the volume in which the radionuclide is contained, and will be mainly high-energy photons. (But note that high-energy β rays contained in a volume with dimensions significantly less than the range of the β rays would also have to be classified as penetrating.)

For non-penetrating radiation the dosimetry is very simple in principle. If all the energy emitted by the radionuclide in the source volume is absorbed in the source volume then the mean absorbed dose, $\bar{D}$, is the total energy emitted divided by the mass of the source volume. Suppose (i) $\tilde{A}$ is the number of nuclear transformations in the source volume during the time interval of interest, i.e. the integral of the activity A in this time interval, (ii) n is the mean number of ionising particles per nuclear transformation, (iii) E is the mean energy per particle and (iv) m is the mass of the volume, then

$$\bar{D} = \tilde{A} \, n \, E/m. \tag{10.7}$$

For penetrating radiation the problem is very much more complex as it is necessary to know not only how much energy is emitted from the source volume but also how much of that energy reaches and is absorbed in those parts of the body where the absorbed dose is required. These volumes can be both inside and outside the source volume. The methods of Marinelli *et al* (1948) were approximate as they did not make accurate allowance for scattered radiation and were normally restricted to geometries which led to mathematical expressions that were readily integrated. Their methods will not be considered here as they have been superseded, but they were employed for about 20 years. Reviews of other developments during that period have been made by Loevinger *et al* (1956) and Loevinger (1969).

These difficult attenuation and scatter problems were resolved by Ellett *et al* (1964, 1965) who used Monte Carlo computer techniques. They introduced the concept of the *absorbed fraction* i.e. the fraction of the particle energy emitted by the source region that is imparted to the volume in which the absorbed dose is required. Following Loevinger and Berman (1968), this latter volume is called the target region. The absorbed fraction is usually represented by the symbol ϕ. (In other chapters of this book the symbol is used for fluence rate.)

If the right-hand side of equation (10.7) is multiplied by the absorbed fraction, ϕ, we have

$$\bar{D} = \tilde{A}\, n\, E\, \phi/m \tag{10.8}$$

and the mean absorbed dose in the target region is obtained for penetrating radiation in an analogous manner to that for non-penetrating radiation. Indeed it can be seen that equation (10.7) for non-penetrating radiation is simply a special case of equation (10.8), having the target region coincident with the source region and an absorbed fraction of unity. The quantity ϕ/m in equation (10.8) is called the specific absorbed fraction and is represented by the symbol Φ. (This symbol is used elsewhere in this book for the quantity fluence.) The product nE of equations (10.7) and (10.8) is the mean energy emitted per nuclear transformation, and is represented by the symbol Δ. Using these symbols equation (10.8) reduces to

$$\bar{D} = \tilde{A}\, \Delta\, \Phi, \tag{10.9}$$

and is of general application as it provides the means of deriving the mean absorbed dose in *any* target region for *any* radionuclide in *any* source region. The reader may feel that equation (10.9) makes the problem look more simple than it is—and the reader would be right. Each of the terms on the right of equation (10.9) presents its difficulties and some of these are considered below.

10.4.4.1 Time integral of activity, $\tilde{A}$, (physiological data). In order to determine $\tilde{A}$ it is necessary to set up a biological model as in principle it is necessary to follow the time course of activity in all parts of the body. Some information is available from animal studies. Indeed, for β-ray emitters, which cannot be detected from outside the body, other sources of data are sparse. The dangers of extrapolation from animals to man are obvious. The distribution of stable elements in man, much of which has been tabulated in ICRP (1975), provides some guidance on the probable sites of concentration of radionuclides in man, but the distribution and retention of an element in the body will depend on the chemical form in which it is introduced. Radioactive daughter products must be considered separately as they may have a different time course through the body. Many measurements have been made on humans,

Table 10.1 Effective dose equivalent from internally administered radionuclides.

Radionuclide	Chemical form	Route of administration	Effective dose equivalent/mSv Activity/MBq
^{3}H	Tritiated water	Oral or i.v.	0.015
^{24}Na	Na$^{\cdot}$	Oral or i.v.	0.4
^{42}K	K$^{\cdot}$	Oral or i.v.	0.2
^{43}K	K$^{\cdot}$	Oral or i.v.	0.2
^{45}Ca	Ca$^{2\cdot}$	Oral	0.6
^{45}Ca	Ca$^{2\cdot}$	i.v.	1.5
^{47}Ca	Ca$^{2\cdot}$	Oral or i.v.	1
^{51}Cr	Labelled normal red cells	i.v.	0.2
^{59}Fe	Fe$^{2\cdot}$ or Fe$^{3\cdot}$	Oral	1
^{59}Fe	Fe$^{2\cdot}$ or Fe$^{3\cdot}$	i.v.	10
^{57}Co	Cyanocobalamin	Oral	2.5
^{58}Co	Cyanocobalamin	Oral	5
^{67}Ga	Ga$^{3\cdot}$	i.v.	0.1

Nuclide	Compound	Route	Value
^{75}Se	Selenomethionine	i.v.	3
^{77}Br	Br	Oral or i.v.	0.7
^{99m}Tc	TcO_4	i.v.	0.01
^{99m}Tc	Human albumin complex	i.v.	0.008
^{99m}Tc	Human albumin {macroaggregates microspheres}	i.v.	0.01
^{99m}Tc	Phosphonate and phosphate compounds	i.v.	0.007
^{99m}Tc	DTPA	i.v.	0.01
^{99m}Tc	Labelled colloid	i.v.	0.01
^{99m}Tc	Labelled normal red cells	i.v.	0.005
^{99m}Tc	Gluconate and Glucoheptonate	i.v.	0.01
^{123}I	I	Oral or i.v.	0.15
^{123}I	Rose bengal or bromsulphthaleine	i.v.	0.08
^{125}I	Iodinated fibrinogen	i.v.	0.1†
^{125}I	Iodinated human albumin	i.v.	{8 / 0.25†}
^{131}I	I	Oral	20
^{201}Tl	Tl	i.v.	0.09

† with thyroid blocked.
From data prepared by the Administration of Radioactive Substances Advisory Committee (UK).

often patients under treatment or investigation with radionuclides, but patients with one disease may not metabolise radiopharmaceuticals in the same way as those who have other diseases or as people who are healthy. Detailed consideration of biological data is outside the scope of this book and reference should be made to ICRU (1979b).

10.4.4.2 Mean energy emitted per nuclear transformation, Δ, (physical data). Suitable tabulations of these physical data exist e.g. Dillman and Von der Lage (1975) and ICRU (1979b). In this latter tabulation Δ is given separately for radiations arranged in five groups, one classed as non-penetrating (β particles $\pm$, electrons, and photons of energy $\leqslant 0.015$ MeV) and four groups of photons of higher energies. It is necessary to include x-rays and conversion electrons in these data, and annihilation radiation following positron emission is added to the photon emission from the radionuclide. ICRP (1983) gives very detailed data on radionuclide transformations in a form useful for dosimetry.

10.4.4.3 Specific absorbed fraction, Φ, (physical and anatomical data). In order to calculate Φ it is necessary to know the energies of the particles emitted by the radionuclide (physical data), the shapes of the source and target regions, the density of these regions, their spatial relationship, and the extent and density of intervening or surrounding tissues (anatomical data). Several sources of data on Φ exist and will be discussed moving from the general to the specific. There is a general reciprocity theorem that can be useful. It states that for any pair of regions in which a radionuclide is uniformly distributed, the specific absorbed fraction is independent of which region is designated the source and which is designated the target. Necessary conditions are either that the regions are situated in an infinite and homogeneous medium or that the medium absorbs the radiation without scatter. The latter condition rarely applies, but the former is approximated to if the two regions are not near the boundaries of the body. Berger (1968) made calculations of specific absorbed fractions for a point source emitting photons of various energies and situated in an infinite volume of water. The specific absorbed fraction is a function of the distance x from the source to the target. It is called the point isotropic specific absorbed fraction and has the symbol $\Phi(x)$. By dividing any target region into a number of smaller regions and weighting $\Phi(x)$ for each of the small regions by the mass of the region, Φ for the whole target region can be obtained. This technique will give some overestimation of Φ in a human being who is, of course, limited in extent, as the infinite water medium assumed by

Berger will ensure that the maximum amount of scattered photons enters the target region. Brownell *et al* (1968) made Monte Carlo calculations which led to values of absorbed fraction for various isotope distributions in a series of cylinders, spheres and ellipsoids. Their results can be interpolated to give absorbed fractions appropriate to anatomical volumes of interest. Snyder *et al* (1969, 1975) have carried out extensive calculations of absorbed fractions in a complex anatomical model which has organs with masses close to those of 'Reference Man', the mythical 'average' 20 to 30 year old weighing 70 kg (ICRP 1975). Although their results will not apply to an actual individual they can be of value in deciding whether more specific calculations are justified.

The results of Snyder and his colleagues have been used by others to estimate the absorbed doses likely to be received following the administration of a number of radiopharmaceuticals. Similar calculations have been based on data developed by the ICRP. For references to many of these works see Appendix E of ICRU (1979b). Selected values of the results of calculations of this kind are shown in table 10.1. The dose quantity used, namely effective dose equivalent (see Chapter 12), is relevant *only* to radiation protection.

11 Methods of Dosimetry

11.1 Calorimeters

So far consideration has been given to three methods of dosimetry, namely calorimetry, ionisation chambers and calculation. Calorimetry is a fundamental method of measuring absorbed dose, but it suffers from being insensitive and by requiring apparatus that normally is complex, is not commercially available, is not readily portable and is slow to operate, as it takes time to reach thermal stability after being set up for use. In consequence calorimetry is normally restricted to standards work or to research applications. (This statement may need modification in the light of the work of Domen (1980, 1982, 1983a, 1983b) who describes a very simple calorimeter.)

11.2 Ionisation Chambers

The principles underlying the use of ionisation chamber dosimetry have been discussed in earlier chapters. Gas-filled ionisation chambers have been used for most of the accurate dosimetry carried out in the world. The associated equipment can be simple and portable and many types are available commercially. By suitable choice of chamber volume, the gas pressure and the sensitivity of the circuit for measuring the ionisation current, an extremely wide range of absorbed dose rates—a factor of more than 10^{16}—can be covered, namely from that of background radiation giving a few times 10^{-8} Gy h^{-1} (a few μrad/h) to water, to that close to the target of an electron accelerator where the absorbed dose rate might be 10^6 Gy s^{-1} (100 Mrad/s).

11.2.1 Saturation corrections

Practical complications in the use of ionisation chambers are few in number. Probably the most important is failure to collect all the ions formed. This can arise in three ways. Of least practical importance is diffusion of ions to the electrode of the same polarity. If the charge carriers are ions and not free electrons the percentage diffusion loss in

124

a parallel plate chamber is approximately 5/number of volts applied (see Greening 1964). Thus even 50 V applied to such a chamber limits diffusion losses to only 0.1%. The second type of loss is through so-called initial recombination, that is recombination between ions formed in a particular track of an ionising particle. For chambers filled with air at atmospheric pressure and used for measurement of photons or electrons, initial recombination is usually insignificant compared to the general (or volume) recombination discussed below (Scott and Greening 1963) but can be significant under the conditions of accuracy required in standards laboratories and where general recombination is reduced by the use of low exposure rates (Niatel 1967). Initial recombination itself is independent of exposure rate as it relates to one particular track and is not affected by the number of other co-existing tracks. If high-pressure gases are used or if the ionising particles are densely ionising, e.g. heavy charged particles or neutrons, initial recombination can be very significant. Theories have been reviewed by Boag (1966). General, or volume, recombination, that is recombination between ions in different tracks, can be very important at high exposure rates. The theory of general recombination has been reviewed by Boag (1966) with further consideration of the special problem associated with pulsed radiation by Boag and Currant (1980) and ICRU (1982). With sufficient attention to detail ionisation chamber measurements can, if necessary, be made to give a precision of better than 0.1%, and this cannot be achieved by other dosimetric methods.

11.3 Chemical Dosimetry

When ionising radiation is absorbed in substances they may be chemically changed, and if this change can be determined it can be used as a measure of the radiation. Hundreds of chemical dosimetry systems have been proposed but few have found application outside the laboratory of initial use. A notable exception is the system proposed over half a century ago by Fricke and his co-workers (e.g. Fricke and Morse 1927). It is called the Fricke or ferrous sulphate dosimeter. In this system, which consists basically of a solution of ferrous sulphate in air saturated dilute sulphuric acid—a more detailed specification is given later—ferrous ions, Fe^{2+}, are oxidised by radiation to ferric ions, Fe^{3+}. The mechanism has been thoroughly investigated and is well understood. The dosimeter solution is usually at least 96% water, and in consequence the radiation interacts almost entirely with the water, ejecting electrons to leave ionised water molecules and also producing water molecules, H_2O^*, excited to a level above the limit of breakage of the H—OH bond. A simplified description of the ensuing reactions is as follows:

$$H_2O \xrightarrow{\text{radiation}} H_2O^+ + e^- + H_2O^*, \qquad (11.1)$$

$$H_2O^* \rightarrow H + OH. \qquad (11.2)$$

A ferrous ion is oxidised to a ferric ion by the hydroxyl radical (another comes from equation (11.7))

$$Fe^{2+} + OH \rightarrow Fe^{3+} + OH^-, \qquad (11.3)$$

and the hydrogen atom from equation (11.2) reacts with dissolved oxygen to give a hydroperoxy radical

$$H + O_2 \rightarrow HO_2, \qquad (11.4)$$

which in turn oxidises a ferrous ion

$$Fe^{2+} + HO_2 \rightarrow Fe^{3+} + HO_2^-, \qquad (11.5)$$

and leads to the formation of hydrogen peroxide

$$HO_2^- + H^+ \rightarrow H_2O_2, \qquad (11.6)$$

which then forms a further ferric ion

$$Fe^{2+} + H_2O_2 \rightarrow Fe^{3+} + OH + OH^-. \qquad (11.7)$$

11.3.1 Radiation chemical yield and *G* value

In chemical dosimetry a quantity, the radiation chemical yield $G(x)$, is introduced which is, for dosimetric purposes, analogous to $1/W$, the reciprocal of the mean energy per ion pair in ionisation chamber dosimetry. (It is perhaps unfortunate that these two methods of dosimetry developed separately. As a consequence we do not have a quantity called the radiation ionisation yield which would have been directly analogous to the radiation chemical yield.) The radiation chemical yield is defined (ICRU 1980) as the quotient $n(x)/\bar{\varepsilon}$ where $n(x)$ is the mean amount of substance of a specified entity, x, produced, destroyed or changed by the mean energy imparted, $\bar{\varepsilon}$, to the matter. Its SI unit is mol J^{-1}. For many years a related quantity called the *G* value has been in use. It is the quotient of the mean *number* (as opposed to *amount of substance*) of elementary entities produced, destroyed or changed and the energy imparted. It has been expressed as the number of entities per 100 eV (i.e. the unit is $(100\text{eV})^{-1}$), and is likely to remain in use until the transition is gradually made to the radiation chemical yield. The two quantities are related by

$$\frac{G\,\text{value}}{(100\,\text{eV})^{-1}} \times 1.0365 \times 10^{-7} = \frac{\text{radiation chemical yield}}{\text{mol J}^{-1}}. \tag{11.8}$$

If the yield of ferric ions can be determined, i.e. if the number of moles can be measured, the energy absorbed can be calculated when the radiation chemical yield is known. Appropriate values for use with the ferrous sulphate dosimeter are given in §11.3.4.

11.3.2 Measurement of absorbed dose

The yield of ferric ions is determined by direct spectrophotometry of the dosimeter solution, which shows absorption peaks in the ultraviolet at 224 and 304 nm. The latter peak is normally used. The absorbed dose, D, in the dosimeter solution is derived from

$$D = \frac{A - A_0}{\rho\,G(\text{Fe}^{3+})\,l\varepsilon_m}, \tag{11.9}$$

where A_0 and A are the absorbance (optical density) of the solution before and after irradiation, ρ is the density of the dosimeter solution, $G(\text{Fe}^{3+})$ is the radiation chemical yield of ferric ion, l is the length of light path in the photometer cell and ε_m is the molar absorption coefficient of ferric ion ($\varepsilon_m = A/lc$ where c is concentration of solute).

Note. If the G value is used instead of the radiation chemical yield, $G(\text{Fe}^{3+})$, the numerator of (11.9) must be multiplied by the Avogadro constant. Also the units in which energy is expressed on the two sides of the equation are likely to be different and a conversion factor will be needed. The derivation of such conversion factors is discussed in §A.1 of the Appendix. If in equation (11.9) SI units are used for all terms on the right-hand side the absorbed dose is obtained in the SI unit Gy.

11.3.3 Practical details of Fricke dosimeter

Chemical dosimetry requires more technical skill than does ionisation chamber dosimetry. Whereas it is possible to purchase an ionisation chamber system and to have it calibrated at a standards laboratory, anyone wishing to use ferrous sulphate dosimetry must prepare pure water, make up solutions with carefully selected chemicals, and ensure the cleanliness and other suitability of all storage vessels and irradiation cells. Also the appropriate value for the molar absorption coefficient of ferric ions needs to be determined for the particular spectrophotometer used for the dosimetry. Some guidance in the necessary procedures is given below.

Nowadays the Fricke dosimeter solution contains about 1 mol m^{-3} of ferrous sulphate or ferrous ammonium sulphate, 1 mol m^{-3} of sodium chloride (to counteract the effects of any organic impurities which may be present despite the precautions taken) and 400 mol m^{-3} of sulphuric acid. The actual quantities required are 0.39 g of ferrous ammonium sulphate (or 0.28 g ferrous sulphate), 0.06 g of sodium chloride and 22 cm^3 of sulphuric acid per litre of final solution. The water should be distilled from alkaline permanganate solution in an all glass or quartz system. The ferrous ammonium sulphate and sodium chloride should be of analytical reagent grade, but the sulphuric acid should be of micro-analytical reagent grade or else should be analytical reagent grade oxidised beforehand to remove trace impurities. The oxidation may be performed with hydrogen peroxide or by irradiation (Davies and Law 1963).

Although pyrex or quartz irradiation cells are preferable for reasons of chemical cleanliness, their use introduces material of atomic number greater than that of water or tissue in contact with the dosimeter solution, giving the boundary problems illustrated in figure 6.3 and discussed in §8.7. Polyethylene cells can be used, and give good homogeneity of the dosimeter system (container plus solution). Plasticised polymers should be avoided (Hall and Oliver 1961), and irradiation cells can with advantage be kept filled with dosimeter solution when not in use. (This solution must be replaced with fresh solution before use.)

The determination of the molar absorption coefficient, ε_m, which occurs in equation (11.9), presents problems. Accurately standardised solutions of ferric ions are difficult to prepare and there was growing evidence (see Svensson and Brahme 1979) that the ε_m value of 220.5 m^2 mol^{-1} at 304 nm and 25 °C recommended by ICRU (1970b, 1972) based on a survey of 83 determinations (Broszkiewicz and Bulhak 1970) may be 1 to 2% high. The ICRU (1982) recommend 216.4 m^2 mol^{-1} based on the work of Eggermont *et al* (1978). Further careful work by this group, reported by Cottens *et al* (1981), leads to a value of 217.4 m^2 mol^{-1}. This point is discussed further in §11.3.4. The molar absorption coefficient increases by 0.7% per °C between 20 and 30 °C (Scharf and Lee 1962).

The linearity of response of the dosimeter should be checked for absorbed doses of 20 Gy (2 krad) to 100 Gy or 150 Gy (10 krad–15 krad). Significant departures from linearity, especially at the lower doses, indicate the presence of impurities. The system should be satisfactory up to 350 Gy using air-saturated dosimeter solution or up to 2 kGy (200 krad) if saturated with oxygen and, for these higher doses, a 4 mol m^{-3} ferrous ion concentration is used. The measurement of pulsed radiation is discussed by ICRU (1982).

The density of the Fricke dosimeter solution when made up with 400 mol m^{-3} sulphuric acid is 1.024 times and its electron density 0.997 times that of water. There is therefore quite close correspondence between the absorbed doses in water and Fricke solution when irradiated with photons of energy such that the Compton effect greatly predominates. As the photon energy falls below about 150 keV the absorbed dose in the Fricke solution will begin to exceed that in water exposed to the same radiation, and becomes some 10–13% higher below 50 keV. The reason for this is the excess absorption in the sulphur of the Fricke solution. The sulphur arises mainly through the sulphuric acid content, and can be reduced by using more dilute acid. The concentration may be reduced to 50 mol m^{-3}, and this reduces the ferric ion yield by only about 2%. Lower acid concentrations should not be used.

11.3.4 Recommended radiation chemical yield and *G* values

G values for the Fricke dosimeter have been determined by many workers. Usually this entails measuring radiation both by the Fricke system and by some other system such as an ionisation chamber or calorimeter. Values for low-energy photons were reviewed by ICRU (1970b) and their recommended values are included in table 11.1. Values for high-energy photons and electrons were also reviewed (ICRU 1969b, 1972) but there were apparent differences between the results of calorimetry on the one hand and ionisation techniques on the other. According to Nahum and Greening (1978) the discrepancies were due to errors in the ionisation dosimetry which were greater for photons than electrons. It now seems appropriate to use the values recommended by ICRU (1969b) for photons up to ^{60}Co energy and to use the ^{60}Co value for all photons and electrons of higher energy. These values are shown in table 11.1. The value for high-energy electrons is supported by the re-evaluation of calorimetric *G* value determinations by Svensson and

Table 11.1 The radiation chemical yield and G values for the ferrous sulphate (Fricke) dosimeter (from ICRU 1969b, 1970b, Nahum and Greening 1978, Svensson and Brahme 1979).

Photon energy†/keV	$G(Fe^{3-})/\mu mol\ J^{-1}$	G value/$(100\ eV)^{-1}$
5	1.30 ± 0.03	12.5 ± 0.3
6	1.32 ± 0.02	12.7 ± 0.2
8	1.35 ± 0.02	13.0 ± 0.2
10	1.37 ± 0.02	13.2 ± 0.2
15	1.41 ± 0.03	13.6 ± 0.3
20	1.43 ± 0.03	13.8 ± 0.3
30	1.46 ± 0.03	14.1 ± 0.3
40	1.48 ± 0.03	14.3 ± 0.3
50	1.49 ± 0.03	14.4 ± 0.3
60	1.50 ± 0.03	14.5 ± 0.3
80	1.51 ± 0.02	14.6 ± 0.2
100	1.52 ± 0.02	14.7 ± 0.2

Radiation		
^{137}Cs	1.59 ± 0.03	15.3 ± 0.3
2 MV	1.60 ± 0.03	15.4 ± 0.3
^{60}Co	1.61 ± 0.02	15.5 ± 0.2
Photons 4 MV–33 MV	1.61 ± 0.03	15.5 ± 0.3
Electrons 3 MeV–35 MeV	1.61 ± 0.03	15.5 ± 0.3

† Problems attend any attempt to attribute a single 'mean' or 'effective' energy to photons having a range of energies (see §1.5). Here it would be appropriate to use the energy of the mono-energetic radiation which has the same half-value layer in aluminium or copper (based on a measurement of exposure) as the radiation in question. Even radiation generated at 250–300 kV is unlikely to have a 'mean' energy much greater than 100 keV. Photons with energies between those generated at 300 kV and those of ^{137}Cs are rarely used.

Brahme (1979) and by the work of Cottens *et al* (1981) and Moses *et al* (1982). Feist (1982) totally absorbed a beam of 5.6 MeV electrons in dosimeter solution and by measuring the charge deposited obtained the total energy transferred to the Fricke solution. His result also agrees with the data in table 11.1. Values of the recently introduced radiation chemical yield, $G(Fe^{3+})$ are also shown.

It will be seen from equation (11.9) that in order to determine absorbed dose it is only necessary to know the product of $G(Fe^{3+})$ and ε_m, and not to know each separately. The agreement between various authors' results is much better in respect of this product than it is for the separate factors. For example, Cottens *et al* (1981), ICRU (1982) and Mosse *et al* (1982) all give values for $G(Fe^{3+})$ and ε_m which lead to a product lying between 348 and 349 $m^2 kg^{-1} Gy^{-1}$. However, ICRU (1984a) recommend 352 $m^2 kg^{-1} Gy^{-1}$, partly because of a problem in knowing the effective density of graphite when calculating its stopping power (of relevance in calorimetric comparisons with ferrous sulphate) and partly because it leads to better agreement between ionisation and calorimetric determinations.

For further information on the ferrous sulphate and other chemical dosimetry systems see Fricke and Hart (1966), and Ellis (1977).

11.4 Thermoluminescence Dosimetry (TLD)

Most crystalline materials when irradiated by ionising radiations store some of the absorbed energy in the crystal lattice. If the materials are subsequently heated some of the stored energy is released as light which can be detected and measured using a photomultiplier. This phenomenon has been known for many years, some materials being artificially activated and studied in the 19th century. The first application to clinical dosimetry appears to have been by Kossel *et al* (1954), but more recent work has been reviewed by Cameron *et al* (1968) and Becker (1973). Thermoluminescence dosimetry differs from the other systems discussed, such as calorimetry, ionisation chambers, and ferrous sulphate dosimetry, in that each crystalline material and read-out system requires calibration against some other dosimeter.

11.4.1 Simple theory of TLD

In an isolated atom the electronic energy states consist of a series of discrete energy levels. In an inorganic crystal lattice, on the other hand, the outer electronic energy levels are perturbed by mutual interactions between the atoms so that the energy levels are broadened into a series of continuous so-called 'allowed' energy bands separated by 'forbidden' energy regions. The bands extend throughout the crystal and the electrons can move within them without additional activation energy. The highest filled band, the valence band, is separated from the lowest unfilled band, the conduction band, by an energy gap of a few eV. If electrons in the valence band receive sufficient energy they can be raised to the conduction band, leaving a vacancy in the valence band called a

positive hole. The electron and positive hole can move independently
through their respective bands. What has been said so far relates to
perfect crystals. In practice variations occur in the energy bands due to
the existence of lattice defects or the presence of impurities. These
variations give rise to local energy levels in the forbidden region between
the valence and conduction bands.

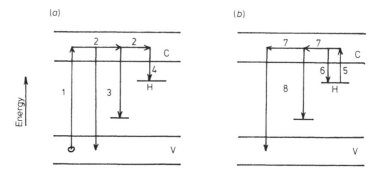

Figure 11.1   The thermoluminescence process.

When the crystal is irradiated an electron in the valence band, V, may
receive enough energy to get raised to the conduction band, C, (figure
11.1a, step 1). It then migrates through the conduction band (step 2)
from which it may fall back to the valence band or higher energy
recombination centre (step 3) or may fall into a trap, H, (step 4). If these
transitions are accompanied by the emission of light the phenomenon is
called fluorescence. If the trap in which the electron is held is not too
deep, the electron may receive enough energy at room temperature to
get out of the trap and recombine in the valence band. In this case the
phenomenon is called phosphorescence. If the trap is deeper the crystal
may need to be raised to higher temperatures before the electron receives
enough thermal energy to be lifted into the conduction band (figure
11.1b, step 5) from where it may fall back into the trap (step 6) or
migrate through the conduction band (step 7) and fall back to a recombination centre or the valence band (step 8). Now if these transitions
are accompanied by the emission of light the phenomenon is called
thermoluminescence. Many processes other than those described above
can take place.

   The deliberate introduction of impurities into crystals increases the
number of traps and can increase the thermoluminescent efficiency of

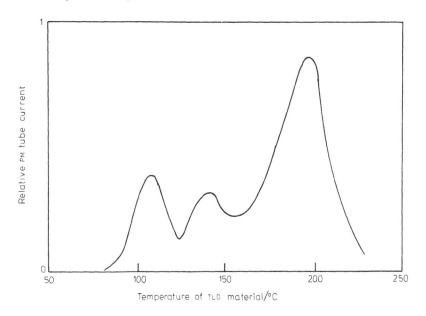

Figure 11.2 Representative glow curve of a TLD material.

the material. Most TLD materials contain these 'activators'. The centres
to which they give rise can have their own characteristic energy levels
and transitions between them give rise to emissions of photons of
corresponding characteristic energies.

As the temperature of a thermoluminescent material that has been
exposed to radiation is increased, more and more trapped electrons are
raised to the conduction band and then emit radiation on falling to lower
energy levels. If the thermoluminescence is plotted against temperature
a glow curve is obtained as illustrated in figure 11.2. The higher temperature part of the glow curve corresponds to electrons in deeper traps.

11.4.2 Dosimetry properties of TLD materials

The materials used for TLD give a measure of the energy absorbed by
the materials themselves and not of the absorbed dose in some other
material such as water or tissue. Let us suppose that it is absorbed dose
in water that is required. If both the water and TLD material are irradiated
by photons under conditions of electronic equilibrium, then the ratio of
the absorbed doses in the two media will be the same as the ratio of

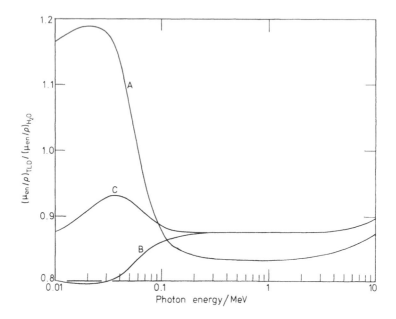

Figure 11.3 Ratio of mass energy absorption coefficients of TLD materials and water. A, LiF; B, $Li_2B_4O_7$; C, $Li_2B_4O_7$ + 0.3% Mn.

their mass energy absorption coefficients. The more the atomic number of the TLD material differs from that of water the greater this ratio will be. It is for this reason that a material such as $CaSO_4$, although having high sensitivity for TLD, has not been used as much for medical dosimetry as materials such as LiF and $Li_2B_4O_7$ which are closer to water or soft tissue in atomic number. Mass energy absorption coefficients for these materials have been given by Hubbell (1982) and their ratios to the corresponding coefficients for water are shown in figure 11.3.

Care is needed in applying these ratios as TLD materials are often exposed under conditions that do not establish electronic equilibrium. These materials can be used as fine powders to a greater or lesser extent surrounded by air or embedded in a plastic matrix (Teflon is most commonly used). The response of the TLD material will then depend on what proportion of the electron energy deposited in it has arisen following photon interactions with the matrix material and what proportion has arisen from photon interactions with the TLD material itself. A difficult case of the 'dosimeter probe' problem discussed in §8.7 exists, and has been discussed by Chan and Burlin (1970).

The response of TLD materials can be affected by, amongst other things, their previous thermal history, their previous radiation history, the time and nature of storage between exposure and read-out, and the gas surrounding them during read-out. Commercial read-out equipment endeavours to control some of these variables. Despite some of these unpromising features TLD has found valuable applications. The great merit of TLD is that the dosimeters can be very small and flexible in shape and can therefore be introduced into materials in which the absorbed dose is required without appreciably perturbing the radiation field. They need no leads or connections and are very rugged. For further information on TLD see the review by Horowitz (1981) and the book by McKinlay (1981).

11.5 Photographic Dosimetry

Photographic films consist of silver halide crystals (mostly silver bromide) embedded in gelatin and spread uniformly and thinly on a thin plastic base. They have been widely employed for personnel dosimetry in radiation protection. The films used for this purpose have had a grain diameter of very roughly 1 μm and emulsion thicknesses of 10–30 μm. Photographic films have also been used for some aspects of dosimetry in radiotherapy, although for these applications thinner emulsions and finer grains have been employed as, for these purposes, it is often advantageous to have a slow film so that exposures may be more accurately controlled.

To a first approximation the radiation-induced optical density of the photographic film is initially found to increase linearly with the exposure to radiation, but eventually falls below the value expected on the basis of direct proportionality. If the film is exposed to photons of different energies and the optical densities lie within the linear range referred to above, it is found that to a first approximation the optical density is proportional to the absorbed dose in the silver halide grains. This quantity is not easy to determine as the silver halide grains receive some energy from electrons released in themselves or their neighbours by the photons, and other energy from electrons released in the gelatin, the film base and other surrounding material. It is a complex 'dosimeter probe' condition, and has been discussed by Greening (1951). It is analogous to the problem of LiF crystals embedded in a plastic matrix referred to in §11.4.2. The theory indicates that the response of a film to radiations of different energy will be influenced by the emulsion thickness and the ratio of silver halide to gelatin in the emulsion. Grain size will also have some effect. Other workers have found grain sensitivity, the presence of additives and activators, and the type of development to influence the response of film to photons of different energies. This variation of response with photon energy (it can amount to a factor of about 40 between 40 and 400 keV) has been a limiting factor in the application of film to dosimetry in radiotherapy, and in dosimetry for radiation protection it has had to be allowed for by assessing the energy of the radiation recorded by the film. This has been done by putting filters of various materials and thicknesses over the film, and using the film itself to measure the attenuations in the filters. After calibration against some other dosimetric system photographic film is capable of giving absorbed doses which are sufficiently accurate for radiation protection purposes. This is not so for applications in radio-therapy, and there photographic dosimetry is used only for measuring *relative* values of absorbed dose. It is also desirable to record the relative values of interest on a single sheet of film so that variations in emulsions from film to film, or of development conditions, are minimised.

Photographic film has some advantageous properties for dosimetric purposes. It can provide a visual representation of the radiation field in the plane in which it lies, or over a curved surface to which it can readily conform. It is valuable, therefore, for determining the position, size and shape of radiation fields. Photographic film has a better spatial resolution than any other dosimetric system and is therefore valuable in making measurements in places where radiation fields are varying rapidly in space, for example close to radiation sources, at the edges of radiation beams and near interfaces, cavities or other heterogeneities.

The film base and gelatin of the photographic emulsion are unlikely to disturb the radiation field in a water- or tissue-equivalent medium, and although the silver halide causes problems in the measurement of low-energy photons it does not cause serious disturbance of the medium when high-energy radiation is being measured, unless the radiation is parallel to the film.

It is possible to record on a single film a great deal of dosimetric information that may form a record for comparison with data obtained on other occasions. All these data can be obtained with little utilisation of time on a machine the main purpose of which is intended not for dosimetric investigation but for radiotherapy. Unfortunately, like chemical dosimetry, photographic dosimetry is rather demanding of good technique.

For further discussion of photographic effects of ionising radiations and film dosimetry see Dudley (1966), Herz (1969) and Becker (1973).

11.6 Scintillation Detectors

Many materials emit light when irradiated with ionising radiations. These materials may be inorganic crystals such as sodium iodide, organic materials such as anthracene, or plastics loaded with appropriate chemicals. The light from these materials can be detected with a photomultiplier tube, and to a first approximation for a particular ionising radiation the output current from the photomultiplier will be proportional to the rate of energy absorption in the detecting material. The scintillator–photomultiplier combination is a very sensitive device and the current output from the photomultiplier might be some 10^8 times as great as that from an ionisation chamber having the same volume as the scintillator. The sensitivity is usually unnecessarily high for measurements in the radiation fields used in radiotherapy, but can be valuable for measurements made for radiation protection purposes.

If photons are to be measured it is usual to calibrate the scintillator in a known radiation field such as that produced by a standardised radium or ^{60}Co source. Alternatively it can be calibrated against an exposure meter providing the exposure meter is sufficiently sensitive for the range of exposure rates it can measure to overlap the range for which the scintillator is used.

If a scintillator is thin enough for the radiation within it to be uniform and yet is thick enough to give full electron build-up when exposed to photons, its response per unit exposure will vary with photon energy approximately as the ratio of the mass energy absorption coefficient of the scintillator material to that of air. The variation of response per unit exposure will therefore be greater for a high atomic number material such as sodium iodide than for one such as a plastic scintillator with an atomic number close to that of air. The energy absorption in the scintillator material will, however, be subject to all the factors discussed in §8.7 which dealt with the theory of the dosimeter probe. The material round the scintillator can affect the response.

If the scintillator has dimensions of a few centimetres, as is common, the position becomes more complex. At low photon energies the scintillator can act as an almost total absorber of the photons, and the mass energy absorption coefficient is irrelevant, except insofar as it is large enough to give total absorption. At somewhat higher energies there will not be total absorption, but (i) the radiation will not be uniform within the scintillator and (ii) some of the secondary photon radiation (Compton scattered photons, characteristic x-rays and annihilation radiation) that is assumed to escape when calculating the energy absorption coefficient will, in fact, be absorbed by the scintillator. Care is therefore necessary in interpreting the readings of a 'thick' scintillation detector when measuring photons of energy other than that for which it has been calibrated.

Similar care is needed when measuring electrons. As long as the electrons lose only a small fraction of their energy in traversing the scintillator, the response will be approximately proportional to the absorbed dose rate in the scintillator material. This absorbed dose rate will in turn be approximately proportional to the absorbed dose rate that would be experienced by a similar thin layer of some other material such as water or tissue if exposed to the electrons, as the stopping power ratio of the scintillator and water is likely to vary only slowly with electron energy. However, as the thickness of the scintillator increases it will absorb more and more of the electrons until it approximates to a total absorption device. It will therefore gradually change from being an indicator of absorbed dose rate to being an indicator of energy fluence rate.

If the scintillator–photomultiplier combination is operated in a pulse counting rather than in a current mode, the response must be interpreted differently. Now every electron with sufficient energy to give rise to a pulse greater than the minimum the discriminator of the counting circuits has been set to accept, will be detected with about 100% efficiency. The device will now be acting as an electron fluence rate detector for all electrons above an energy corresponding to the discriminator setting.

For more detailed discussion of scintillation dosemeters see Ramm (1966).

11.7 Other Dosimetric Systems

Many other dosimetric systems have been suggested and brief mention will be made of a selected few.

Geiger counters have some similarities to scintillation detectors used in the pulse count rate mode. They have an efficiency of nearly 100% for electrons (β rays) which are sufficiently energetic to penetrate the wall of the counter and enter the sensitive volume of the counter. However, for photons their efficiency is low—of the order of 1%—as they only respond to photons absorbed in the thin layer of the wall from which the ejected electrons can reach the sensitive volume. Their use is largely restricted to radiation protection measurements, but they are sometimes used as the neutron insensitive detector in the double dosimeter method that is employed to measure the mixed neutron and γ ray field experienced in neutron therapy (see §9.4). Further discussion of Geiger counters has been provided by Emery (1966).

Silicon can be used as a solid state ionisation chamber (Parker and Morely 1966, Fowler 1966). Volume for volume it is over 10^4 times as sensitive as a gas-filled ionisation chamber. A factor of about 10^3 comes from the ratio of densities and another order of magnitude comes from the energy per electron–hole pair in silicon irradiated by photons or electrons being only 3.68 eV at 300 K compared with 33.8 eV for air. Silicon is closer to bone than to soft tissue in its interactions with photons, but its stopping power for electrons relative to that of water varies little with energy, and it may therefore have more application as a dosimeter for high-energy electrons.

The ultraviolet absorption of various plastics is changed by irradiation and this phenomenon can be applied to dosimetry. Polymethyl methacrylate (Perspex, Lucite) has been used by Boag *et al* (1958) and Orton (1966) among others, and polyethylene terephthalate (Mylar, Melinex) by Ritz (1961). Colour changes are observed on the irradiation of dyed polymers, and Whittaker (1970) has used red Perspex for dosimetry. These polymers are a close match dosimetrically speaking to materials of biological interest, and are of value in measuring the large absorbed doses required for radiation sterilisation of materials. Becker (1973) has provided a useful review of these dosimetric systems.

Very high absorbed dose rates and large pulses of radiation can be measured using vacuum chambers, that is devices like ionisation chambers from which the gas has been removed. The current passing through such chambers arises, not from ionisation, but from an emission of low-energy electrons from the chamber electrodes. These electrons are of sufficiently low energy to be influenced by the potential difference applied between the electrodes (Greening 1954) and as no recombination can occur large absorbed dose rates or pulses can be measured. The subject has been reviewed by Burlin (1972).

11.8 Choice of Dosimetric System

Some of the criteria which govern the choice of a dosimetric system are set out in table 11.2. Often the application of one or two of these criteria severely limits the choice of system. Thus if an absolute method is required only calorimetry or an ionisation method is applicable. If, in addition, the absorbed dose to be measured is less than about 1 Gy (100 rad) only the ionisation method remains (see figure 11.4). A need to measure less than 10^{-6} Gy (10^{-4} rad) restricts the choice to ionisation or scintillation methods (figure 11.4). If neutrons are to be measured accurately it is necessary to have particularly close matching of the dosimetric material to that in which the absorbed dose is required. Personnel monitoring with its requirement for a small, rugged, low-cost, integrating device able to measure low absorbed doses, but without a need for high accuracy, soon brings the choice down to photographic methods or TLD. In general, absorbed dose rate requirements put fewer restrictions on the choice of dosimetric systems (figure 11.5). However,

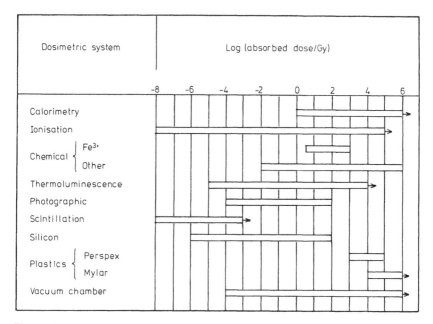

Figure 11.4 Approximate range of absorbed doses measurable with different dosimetric systems (modified from Boag 1967).

Table 11.2 Some criteria governing the choice of a dosimetric system (extended from Boag (1967)).

Whether an absolute method is required
Accuracy and precision required
Whether total absorbed dose or absorbed dose rate is required
Whether an immediate read-out is required
Range of absorbed dose to be measured
Range of absorbed dose rate to be measured
Type and energy of radiation to be measured
Need to match dosimeter to medium
Size of detector required
Spatial resolution required
Convenience
Cost
Ruggedness

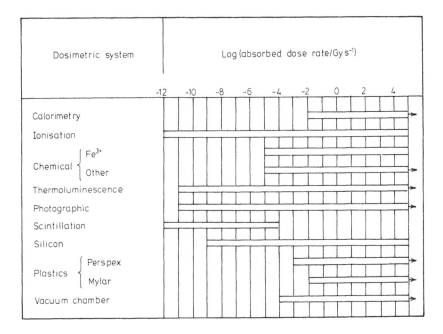

Figure 11.5 Approximate range of absorbed dose rates measurable with different dosimetric systems (modified from Boag 1967).

calorimetry normally requires more than about 10^{-2} Gy s^{-1} (1 rad s^{-1}) if the radiation heating is to be substantially greater than the heat exchange between the thermal element and its surroundings. Also those systems requiring high absorbed doses are unlikely to be chosen for the measurement of low absorbed dose rates even if, in principle, they could be utilised, because of the time required for a measurement.

12 Dosimetry in Radiation Protection

12.1 Dose Equivalent

Unwanted biological effects of ionising radiation were recorded in January 1896 only two months after Roentgen's discovery of x-rays. Sporadic measures were adopted to reduce these radiation hazards but the first *organised* steps appear to have been those of the British Roentgen Society in 1915. Even then progress was slow as any recommendations as to the maximum amounts of radiation which people could safely receive had to await, among other things, a scientifically sound quantity and unit in which the radiation could be expressed. The body now known as the International Commission on Radiological Protection was set up in 1928 during the Second International Congress of Radiology, and its first recommendations on 'dose' levels of radiation were made in 1934 in terms of the 'roentgen' which had been defined by the ICRU in 1928 (see §5.1). The recommendations related only to x-rays and, although the use of the roentgen was extended to γ-ray measurements by the ICRU in 1937, the ICRP did not make recommendations at that time for radium γ rays.

During the war years a great deal of radiobiological data was accumulated, and in the 1950s consideration was given by the ICRP to neutrons, protons and other heavy charged particles which were by then more prevalent following the development of nuclear reactors and particle accelerators. But these radiations could not be expressed in terms of roentgens, which are usable only with photons. However, it was possible to calculate the energy absorbed per unit mass of soft tissue when exposed to one roentgen of x- or γ radiation, and this energy per unit mass—within a few percent of what was shortly to be known as the rad—was employed for the neutrons, protons and heavy charged particles. It was also recognised that these particles laid down energy much more densely along their tracks than did electrons, and in consequence produced different biological effects for a given energy deposition per unit mass. Selected values of this relative biological effectiveness (RBE) of

143

the densely ionising radiations were multiplied by the energy per unit mass referred to above to give a quantity called the 'RBE dose' which was expressed in the unit 'rem' which at that time meant 'roentgen equivalent man'. Shortly after this the quantity absorbed dose was introduced with its unit the rad (see Chapter 4) and this permitted all radiations to be measured in terms of the same quantity and unit. The ICRU (1962) introduced a quantity called dose equivalent, H, to replace the old RBE dose. Initially the dimensions of the quantity were not indicated and its meaning was interpreted by different people in different ways, but it is now seen to be (ICRU 1980) a weighted absorbed dose given by

$$H = D \, Q \, N, \tag{12.1}$$

where D is the absorbed dose, Q is one weighting factor, the quality factor, and N is the product of all other weighting factors that might modify the potentially harmful biological effects of the absorbed dose D.

12.2   Quality Factor, Q

The quality factor is conceptually related to the RBE factor of the 1950s. The reason the old RBE factor was replaced by the quality factor is as follows. The biological effectiveness of an absorbed dose of a particular radiation relative to that of a standard radiation is normally experimentally observable by subjecting a particular biological system to both radiations in turn and, while keeping all other factors constant, finding the ratio of the absorbed doses required to reach the same biological end point with the two radiations. In radiation protection many different organs are involved and many different biological effects. Furthermore, it is rarely possible to carry out an experiment that gives significant results at the very low absorbed doses with which we are concerned in radiation protection, and the effects at low doses are inferred by judgmental extrapolation (see below) from experiments employing higher absorbed doses. It was therefore thought undesirable to use the scientific term RBE for the assigned weighting factor covering all deleterious biological effects at low doses. The quality factor, Q, is specified in terms of the linear collision stopping power, S_{col}, in water, which is equal to the unrestricted linear energy transfer, L_∞ (see Chapter 2). The relationship between Q and L_∞ is specified by the ICRP (1977) and is shown in table 12.1.

Both ICRU and ICRP documents speak of the quality factor being 'a

Table 12.1 Relationship between Q and L_∞ in water, specified by ICRP (1977).

L_∞ in water/keV μm^{-1}	Q
3.5 (and less)	1
7	2
23	5
53	10
175 (and above)	20

function of' the unrestricted linear energy transfer. It should be made clear that it is not a function in the sense that given L_∞ it is possible to calculate Q from first principles. Initially the biological effects of a radiation of a particular L_∞ in water were compared with those of photon radiation (L_∞ in water less than $3.5\,\text{keV}\,\mu m^{-1}$) and RBE values were determined. These RBE values were extrapolated in what was thought at the time to be an appropriate manner to the low dose levels of relevance to radiation protection. Then a value judged to be representative of the various biological effects which give concern at low doses was chosen. The procedure was repeated for radiations of different L_∞ in water. The values so derived are now called Q. They have remained unchanged since they first appeared in the 1950s as 'RBE factors', despite the increase in radiobiological theory and experimental data since that time. Newcomers to the field should not read more into the numbers than is justified by the original data. Do not be misled by the use of two- or even three-figure numbers in column 1 of table 12.1, but observe that the values of Q in column 2 mostly change by a factor of 2. Because of the way in which they have been derived, values of Q are only applicable for routine radiation protection purposes and should not be used in assessing the effects of accidental doses substantially above the limits laid down by ICRP.

The great majority of protection measurements involve only photons or electrons, and for these Q is unity. If other or mixed radiations are to be measured a range of values of Q arises and a mean value, $\bar{Q}$, weighted by absorbed dose is employed. This is given by

$$\bar{Q} = \frac{1}{D}\int_0^\infty Q\,D_{L\infty}\,dL_\infty, \qquad (12.2)$$

where $D_{L\infty}\,dL_\infty$ is the absorbed dose deposited by radiation having linear energy transfer lying between L_∞ and $L_\infty + dL_\infty$.

12.3   Product, *N*, of Other Weighting Factors

The deleterious biological effects of ionising radiations may be affected by other factors such as the distribution of the radiation in time, but in the state of present knowledge the product of all weighting factors other than *Q* has been assigned the value 1 by the ICRP.

12.4   The Unit of Dose Equivalent

As both *Q* and *N* are pure numbers, both *H* and *D* have the same dimensions. The SI unit of both is the same, namely $J\,kg^{-1}$. Whereas this unit is given the special name gray, Gy, when used for absorbed dose it is given the special name sievert, Sv, when used for dose equivalent. This rather unusual arrangement has been permitted by the General Conference on Weights and Measures to help avoid the hazards that could arise if absorbed dose were confused with dose equivalent. Unfortunately the name of both quantities contains the word 'dose', and both are frequently contracted to 'dose' not only in speech but in writing. The misunderstandings that arise over these two quantities and their units are discussed further in §A.2 of the Appendix.

The special unit of dose equivalent, the rem, can be used temporarily but, like the other special radiation units, roentgen, rad and curie, is due to be phased out by about 1985. In present day terms the rem is $10^{-2}\,J\,kg^{-1}$, the old connection with the roentgen having been dropped when the quantity dose equivalent was introduced in 1962.

12.5   Effective Dose Equivalent

The ICRP (1977, 1978) introduced a number of additional quantities for use in radiation protection, one of which will be considered here. The limits ICRP set for dose equivalent are based on an assumption of uniform irradiation of the body. In practice this is very rare. Moreover different tissues have different susceptibilities to radiation. In order to take these two considerations into account ICRP defined *effective dose equivalent*, H_E, as

$$H_E = \Sigma w_T H_T, \qquad (12.3)$$

where w_T is a weighting factor representing the proportion of the total risk arising from uniform irradiation of the whole body which is to be

Table 12.2 Tissue weighting factors, w_T, recommended by
ICRP (1977).

Tissue	w_T
Gonads	0.25
Breast	0.15
Red bone marrow	0.12
Lung	0.12
Thyroid	0.03
Bone surfaces	0.03
Remainder of body†	0.30

† Excluding hands and forearms, feet and ankles, the skin and
the lens of the eye (ICRP 1978).

attributed to irradiation of tissue T, and H_T is the dose equivalent in
tissue T. The values of w_T recommended by the ICRP are shown in table
12.2. The unit of effective dose equivalent is the same as that of dose
equivalent, namely the sievert or, temporarily, the rem.

12.6 Determination of Dose Equivalent

It will be seen from equation (12.1) that dose equivalent, H, is obtained
by determining absorbed dose, D, and multiplying by the appropriate
weighting factors, Q and N. Fundamentally, therefore, dosimetry for
radiation protection purposes is no different from that for any other
purpose. Indeed, as most protection measurements involve only photons
and electrons, for which Q and N are both unity, all that is usually
required is a determination of absorbed dose. In practice there are
special problems.

First, the absorbed dose rates are often very low. In consequence
sensitive instruments are required such as large volume ionisation chambers, scintillation detectors or geiger counters. The interpretation of the
readings of these last two devices has been discussed briefly in §11.6 and
§11.7 respectively. Second, the small absorbed dose rates may need to
be integrated over extended periods. This leads to the use of photographic films and TLD as discussed in Chapter 11. Third, the assessment of
absorbed doses arising from unsealed radioactive materials accidentally
taken into the body is always based on incomplete data. Even when the

radionuclide, its activity, its chemical form and its time and route of introduction into the body are known, as is the case with the medical use of radionuclides discussed in Chapter 10, absorbed dose determination is hard enough. When any or all of these data are in doubt, the problem is even less amenable to accurate solution. Fourth, it is not possible to measure the separate absorbed doses to all the parts of the body mentioned in table 12.2. Fifth, unlike the situation in radiotherapy where the size and orientation of the radiation beam is known, in protection measurements the radiation 'beam' may have varied and unknown size and direction. For these last two reasons the radiation may be measured over a period at one or more sites on the body and then the absorbed doses may be inferred at points within the body. Sixth, it is often necessary to make measurements in unoccupied places in order to determine what absorbed doses people would receive were they subsequently to be at those places (see §§12.7 and 12.8).

Because of the special difficulties of absorbed dose measurements in radiation protection, and because of the uses to which they are put (for example, checking that regulations are complied with or assessing the efficacy of additional protective measures) many simplifications and approximations are acceptable. The most common simplification is to assess the *maximum* absorbed dose or dose equivalent occurring in the body. If this is satisfactory more details of absorbed dose distribution are rarely necessary. Under most circumstances the maximum dose equivalent occurs at very nearly the same depth in the body as the maximum absorbed dose. Indeed with photons and electrons for which Q is 1 the maxima must coincide. For neutrons of unknown energy, the adoption of a value 10 for Q will ensure that H is not underestimated. (For an alternative approach see §12.10.)

12.7   The Index Quantities

The introduction of a person into a radiation field may well alter that field through the absorption and scattering of radiation by the human body. Indeed, because of this scattering the exposure at a point in a radiation field of 100 keV photons could well be increased by over 50% on the entry of a person into the radiation field. It can be argued that when investigating a radiation field for protection purposes it is better to measure a quantity that is representative of the situation *after* the introduction of a human body. This was one of the reasons that led the

ICRU (1971a) to define two new quantities, absorbed dose index, D_I, and dose equivalent index, H_I. These are respectively the highest absorbed dose and the highest dose equivalent that occur in a 30 cm diameter sphere of unit-density tissue-equivalent material centred at the point of interest. The 30 cm is an approximation to the dimension of a human body that is significant in the attenuation and scattering of radiation and the spherical shape renders the determination of the index quantities directionally independent.

The index quantities have some unusual properties. They rarely occur at the point at which they are said to exist, namely the centre of the 30 cm sphere. Also in a mixed radiation field the positions in the sphere of the maximum absorbed doses for the individual radiations are unlikely to coincide. In consequence the overall absorbed dose index is nearly always less than the sum of the individual absorbed dose indices, and is never more than that sum. A similar argument holds for the separate and overall absorbed dose indices for several beams of the same radiation coming from different directions.

The ICRU (1976) extended the concept of the dose equivalent index by proposing two restricted dose equivalent indices, the shallow dose equivalent index, $H_{I,s}$, and the deep dose equivalent index, $H_{I,d}$. The former, $H_{I,s}$, is the maximum dose equivalent occurring in the outer 1 cm 'rind' of the 30 cm sphere when the outermost 70 μm is ignored, and $H_{I,d}$ is the maximum dose equivalent in the remaining 28 cm diameter 'core'. The 70 μm is chosen as being a representative depth for the basal layer of the skin, and radiation effects in more superficial cells are assumed to be negligible.

12.8 Practical Quantities

Several facts pointed to a need for supplementary new quantities amenable to practical measurement. First, the index quantities had some undesirable properties and had not been widely adopted in radiation protection. As a consequence different radiations often continued to be measured in terms of different quantities. Thus photons might be measured as exposure expressed in roentgens, electrons and β rays as kerma or absorbed dose in air or soft tissue expressed in rads, and neutrons as dose equivalent expressed in rems. The gradual change from special radiation units (roentgen, rad and rem) to SI units highlighted—but did not cause—this untidy situation, as the approximate numerical equivalence between exposure (in roentgens) and absorbed dose to water or tissue (in rads) and hence to dose equivalent (in rems), disappeared when the respective SI units $C kg^{-1}$, Gy and Sv were used (see §7.3). Exposure meters needed to be rescaled in terms of $C kg^{-1}$, or, for protection purposes, recalibrated in terms of an agreed and relevant protection quantity. Second, statements of dose equivalent were usually vague as to the tissue or tissues that had received the radiation. Third, although the introduction in 1977 by the ICRP of the quantity effective dose equivalent had defined many (but not all) of the tissues to be considered (table 12.2), the quantity was virtually unmeasurable. Quantities needed to be introduced that were applicable to all radiations and were readily measurable by fairly simple equipment that could have calibrations traceable to national standards. Furthermore, these quantities must have a known quantitative relationship to effective dose equivalent and to the mean dose equivalent in organs, as radiation protection regulations are now specified in terms of those concepts.

Very extensive Monte Carlo calculations were made on behalf of ICRU (ICRU, to be published) of the distribution of dose equivalent in the trunk of an anthropomorphic phantom and in the ICRU 30 cm diameter sphere, when irradiated by photons, neutrons and electrons of various energies. It was found that the effective dose equivalent and organ doses in the phantom could be adequately estimated (with very rare underestimation and without excessive overestimation) from the dose equivalent at certain specified depths in the ICRU sphere irradiated by the same radiation (see §12.9 and figures 12.1 and 12.2). It is therefore possible for appropriate instruments to be calibrated at standardising laboratories against the dose equivalent at specified depths in the ICRU sphere. These instruments can then be used for measurements in the field to get estimates of the quantities in terms of which ICRP makes its recommendations.

On the basis of the above considerations ICRU (1985) defined additional quantities for use in radiation protection. Some preliminary definitions were necessary. If a measurement is made at only one point in a radiation field that will later be occupied by a person, it is necessary to assume that the field at that point extends over the volume of the person. Thus ICRU defined an *expanded field* as one in which the fluence and its angular and energy distributions are the same throughout the volume of interest as in the actual field at the reference point. Also in an *aligned and expanded field* the fluence and its energy distribution are the same as in the expanded field but the fluence is unidirectional.

12.8.1 Environmental monitoring

Two quantities were defined for environmental monitoring—one for strongly penetrating radiation and another for weakly penetrating radiation.

12.8.1.1 Ambient dose equivalent, $H^*(d)$. This quantity is recommended for use with measurements of strongly penetrating radiation. The ambient dose equivalent, $H^*(d)$, at a point in a radiation field is the dose equivalent that would be produced by the corresponding aligned and expanded field, in the ICRU sphere at a depth d on the radius opposing the direction of the aligned field. It is recommended that $d = 10$ mm and then $H^*(d)$ may be written as $H^*(10)$.

The ideal instrument for measuring H^* will have an isotropic response as it will measure H^* in any radiation field that is uniform over the dimensions of the instrument.

12.8.1.2 Directional dose equivalent, $H'(d)$. This quantity is recommended for use with weakly penetrating radiation (for example where the skin is likely to be the organ of greatest interest). The directional dose equivalent, $H'(d)$, at a point in a radiation field, is the dose equivalent that would be produced by the corresponding expanded field in the ICRU sphere at a depth d on a radius in a specified direction. The recommended depth for environmental monitoring in terms of $H'(d)$ is 0.07 mm (the conventional depth for the basal layer of cells in the skin), and $H'(d)$ may then be written as $H'(0.07)$.

Instruments for the measurement of this quantity should accept radiation from a solid angle of 2π sr. There should be enough material behind the sensing element to provide backscatter of radiation from the solid angle of 2π sr mentioned above, and attenuation of radiation from the remaining 2π sr, i.e. radiation incident on the rear of the instrument.

12.8.2 Individual monitoring

Two further concepts were introduced by ICRU (1985) for purposes of individual monitoring, one for estimating appropriate dose equivalents to deep organs irradiated by strongly penetrating radiation and another for use when shallow organs and weakly penetrating radiation are involved.

12.8.2.1 Individual dose equivalent, penetrating, $H_p(d)$. The individual dose equivalent, penetrating, $H_p(d)$, is the dose equivalent in soft tissue below a specified point on the body at a depth, d, that is appropriate for strongly penetrating radiation. The quantity can be measured with a detector worn on the surface of the body and covered with an appropriate thickness of tissue equivalent or substitute material.

The recommended depth, d, is 10 mm and $H_p(d)$ may then be written $H_p(10)$. Dosemeters should be calibrated on an appropriate phantom. For dosemeters worn on the trunk an ICRU sphere is suitable.

The quantity $H_p(10)$ at a given location on the anterior of the trunk can be related to the effective dose equivalent, H_E, received by the trunk when irradiated from front and sides.

12.8.2.2 Individual dose equivalent, superficial, $H_s(d)$. The individual dose equivalent, superficial, $H_s(d)$, is the dose equivalent at a specified point on the surface of the body under a layer of thickness d. The recommended depth, d, is 0.07 mm and $H_s(d)$ may then be written $H_s(0.07)$. This quantity can be measured with a detector on the surface of the body covered by 0.07 mm of tissue equivalent, or suitable substitute, material. Again calibrations are carried out with dosemeters on a suitable phantom, namely one that is of a size and shape that sufficiently represents the part of the body on which the dosemeter will be worn.

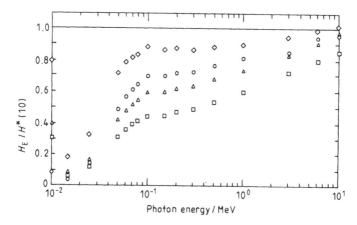

Figure 12.1 The ratio of effective dose equivalent H_E, for different orientations of an anthropomorphic phantom, to the ambient dose equivalent, $H^*(10)$, when exposed to the same beam, as a function of radiation energy. ◇ Anterior–posterior irradiation of anthropomorphic phantom, ○ posterior–anterior irradiation of anthropomorphic phantom, △ phantom rotated about long axis, □ isotropic irradiation of anthropomorphic phantom. Photons (Williams *et al* 1983).

12.9 Relationships between Monitoring and Regulatory Quantities

Figures 12.1 and 12.2 give the ratio of effective dose equivalent, H_E, to ambient dose equivalent $H^*(10)$, for photons and neutrons, respectively, over a wide energy range. The results of Monte Carlo calculations for various irradiation conditions are illustrated, namely AP, PA, rotational and isotropic fields applied to an anthropomorphic phantom. It will be seen that in all cases the ratio is less than unity so that $H^*(10)$ over-estimates H_E. Thus if $H^*(10)$ does not exceed the limits for effective dose equivalent laid down by ICRP, neither will H_E.

$H^*(10)$ gives an appreciable overestimation of H_E for low energy photons and neutrons with energies around 10^5 eV. If this is unacceptable more complex measurements will be necessary.

Figures 12.1 and 12.2 are representative of many available to ICRU which show that individual organ doses also are not underestimated by $H^*(10)$ (or by $H'(0.07)$ in the case of skin).

12.10 Dose Equivalent Measurements of Neutrons

If the dose equivalent resulting from neutron irradiation is to be determined from a measurement of absorbed dose it is necessary to know the

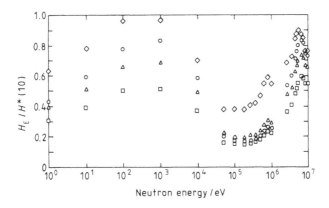

Figure 12.2 Ratio of H_E to $H^*(10)$ for neutrons (Burger *et al* 1984). For explanation of symbols, see caption to figure 12.1.

appropriate $\bar{Q}$ specified in equation (12.2). However this necessity can be avoided by obtaining a direct indication of the maximum dose equivalent in the body. This has been done by designing instruments the response of which varies with neutron energy in the same way as the factor converting neutron fluence to maximum dose equivalent varies with neutron energy. These instruments have been called rem-meters and their design and properties have been reviewed by Nachtigall and Burger (1972).

12.11   Sources of Data on Radiation Protection Instrumentation and Methods

This chapter has concentrated on the fundamental aspects of dosimetry in radiation protection, but other information will be required by those making practical measurements in this field. The literature of the subject is vast but a good entry into it will be provided by ICRU (1971b), HPA (1973) and NCRP (1978).

Appendix: Physical Quantities and Units

A.1 Equations

A short section on physical quantities and units might appear to be out of place in a book on dosimetry. However, the discussions associated with the introduction of SI units into dosimetry and the phasing out of the special radiation units have revealed a surprisingly widespread confusion about the subject.

The value of a physical quantity is equal to the product of a numerical value (a pure number) and a unit:

$$\text{physical quantity} = \text{numerical value} \times \text{unit}. \qquad (A.1)$$

As the numerical value is dimensionless the physical quantity and the unit must have the same dimensions. Indeed a unit is a selected reference sample of a physical quantity. This puts restrictions on the units that may be used in conjunction with a particular physical quantity, but reference to a physical quantity should not imply any particular choice of unit. A physical quantity is the same no matter what the units in which it is expressed. The *length* of a metal bar is clearly the same whether it is expressed in inches or metres. A particular *absorbed dose* is the same whether it is expressed in rads or in grays. The *numerical value* will, of course, be one hundred times greater when the unit is the rad than when it is the gray, since 1 gray = 100 rad. Thus an equation which gives the relationship between physical quantities is true no matter what units may subsequently be employed, and no units have to be stated. Physical quantities have usually been defined in such a way that the relationships between them do not contain numerical factors. Thus in

$$\text{velocity} = k \, \frac{\text{distance}}{\text{time}}, \qquad (A.2)$$

k has been assigned the value unity. However, the factor 4π sometimes

155

occurs in relationships between quantities, and occasionally putting $k = 1$ in a series of such relationships leads to a numerical factor in a relationship between derived quantities. The most common example is probably

$$\text{kinetic energy} = \tfrac{1}{2} \text{mass} \times (\text{velocity})^2. \qquad (A.3)$$

These exceptions apart, the appearance of a number in an equation implies that numerical values have been assigned to quantities and this means that units have been chosen. Such equations with numbers need to have the units stated, as they are only true for the particular units used. The moment particular units are introduced into a relationship between physical quantities, constants of proportionality may occur. Thus suppose in the equation

$$\text{velocity} = \frac{\text{distance}}{\text{time}}, \qquad (A.4)$$

we express velocity in centimetres per second, distance (rather perversely) in inches and time in seconds. If we divide each quantity by the unit in which it is expressed we will get a relationship between numerical values. This will only be true if a factor is introduced which gives the relative magnitudes of the units concerned. Thus

$$\frac{\text{velocity}}{\text{cm s}^{-1}} = \frac{\text{distance}}{\text{in}} \times \frac{\text{s}}{\text{time}} \times \frac{\text{in}}{\text{cm s}^{-1} \text{ s}}. \qquad (A.5)$$

All the units have been gathered together in the final term on the right-hand side in such a way as to cancel the units in the separate terms. If the equation is dimensionally correct this term will be dimensionless. Although at first sight it may look pedantic to divide each quantity by its unit it is an excellent method of checking that the equations are indeed dimensionally correct. In general, many units in terms such as the last in equation (A.5) will cancel, but if different units have been employed for quantities of the same dimensions (albeit often in a much more concealed manner than in equation (A.5)), we will finish up with a ratio of units of the same dimension, such as in/cm, and must use the relationship between them, namely 1 in = 2.54 cm, giving a final equation

$$\text{velocity} = 2.54 \, \frac{\{\text{distance}\}}{\{\text{time}\}} \text{ cm s}^{-1}, \qquad (A.6)$$

where {distance} is the numerical value when distance is expressed in inches and {time} is the numerical value when time is expressed in seconds.

If a coherent set of units, such as SI, is used, constants of proportionality are unity. Thus

$$\frac{\text{velocity}}{\text{m s}^{-1}} = \frac{\text{distance}}{\text{m}} \times \frac{\text{s}}{\text{time}} \times \frac{\text{m}}{\text{m s}^{-1}\,\text{s}}, \tag{A.7}$$

and the final term on the right-hand side is unity,

$$\text{i.e. velocity} = \frac{\{\text{distance}\}}{\{\text{time}\}}\ \text{m s}^{-1}, \tag{A.8}$$

where {distance} is the numerical value when distance is expressed in metres and {time} is the numerical value when time is expressed in seconds.

To take a dosimetric example let us reproduce equation (7.2),

$$D_{\text{m}} = X\,\frac{W_{\text{air}}}{e}\,\frac{(\mu_{\text{cn}}/\rho)_{\text{m}}}{(\mu_{\text{cn}}/\rho)_{\text{air}}}. \tag{A.9}$$

If each quantity is divided by its SI unit we obtain

$$\frac{D_{\text{m}}}{\text{Gy}} = \frac{X}{\text{C kg}^{-1}}\,\frac{W_{\text{air}}/e}{\text{J C}^{-1}}\,\frac{(\mu_{\text{cn}}/\rho)_{\text{m}}}{(\mu_{\text{cn}}/\rho)_{\text{air}}}\,\frac{\text{C kg}^{-1}\,\text{J C}^{-1}}{\text{Gy}}. \tag{A.10}$$

(It is assumed that the same unit will be used for both the μ_{cn}/ρ, i.e. that the ratio of the units is unity.) As $\text{Gy} = \text{J kg}^{-1}$ the final term is unity. Inserting the known value of 33.85 J C^{-1} for W_{air}/e we have

$$D_{\text{m}} = 33.85\,\{X\}\,\frac{(\mu_{\text{en}}/\rho)_{\text{m}}}{(\mu_{\text{en}}/\rho)_{\text{air}}}\ \text{Gy}, \tag{A.11}$$

where {X} is the numerical value when exposure is expressed in C kg^{-1}. But using the special radiation units, rad and roentgen, together with the electronvolt, we obtain

$$\frac{D_{\text{m}}}{\text{rad}} = \frac{X}{\text{R}}\,\frac{W_{\text{air}}}{\text{eV}}\,\frac{\text{C}}{e}\,\frac{(\mu_{\text{en}}/\rho)_{\text{m}}}{(\mu_{\text{en}}/\rho)_{\text{air}}}\,\frac{\text{R eV}}{\text{C rad}}. \tag{A.12}$$

The final term is

$$\frac{2.58 \times 10^{-4} \text{ C kg}^{-1} \times 1.602 \times 10^{-19} \text{ J}}{\text{C } 10^{-2} \text{ J kg}^{-1}},$$

i.e. $2.58 \times 1.602 \times 10^{-21}$.

This is a pure number and the equation is therefore dimensionally correct. If we also insert the known values of $W_{air} = 33.85$ eV, and of the electronic charge, $e = 1.602 \times 10^{-19}$ C we have

$$D_m = \frac{2.58 \times 1.602 \times 10^{-21} \times 33.85}{1.602 \times 10^{-19}} \{X\} \frac{(\mu_{en}/\rho)_m}{(\mu_{en}/\rho)_{air}} \text{ rad},$$

$$= 0.873 \{X\} \frac{(\mu_{en}/\rho)_m}{(\mu_{en}/\rho)_{air}} \text{ rad}, \qquad (A13)$$

where $\{X\}$ is the numerical value when exposure is expressed in R.

In equation (A.5) it may have seemed perverse to express velocity in cm s^{-1} and yet to express distance in inches. This is because it is readily appreciated that velocity includes the quantity length. But in equation (A.12) it is not so apparent that the energy 'hidden' in D_m is not being expressed in the same units as the energy of W_{air}. Neither is it obvious that the charge involved in the unit R is a sub-multiple of the coulomb.

A.2   Absorbed Dose and Dose Equivalent

The quantities and units used in radiation protection seem to cause particular problems. We have the following relationship between quantities:

dose equivalent = absorbed dose × quality factor × other factors,

$$H = D \, Q \, N.$$

As the 'other factors' are currently assigned the value 1 let us consider H, D and Q, and suppose we are concerned with particular neutrons for which $Q = 10$. As Q is a pure number the dimensions of H must be the same as those of D. The SI unit for D is J kg^{-1} (given the special name, gray), and the SI unit for H is also J kg^{-1} (but now given the different special name, sievert). In the above example an absorbed dose of 10^{-3} J kg^{-1} would give rise to a dose equivalent of 10^{-2} J kg^{-1}. Many people find this confusing. The essential point is that absorbed dose and dose equivalent are *different quantities* although they have the same dimensions and are related by the non-dimensional factor Q. Similarly the diameter of a circle and its circumference are different quantities of the same dimension related by the non-dimensional factor π, and a diameter of 1 cm implies a circumference of 3.14 cm. In attributing a quantity of 10 cm to a circle one would expect to say whether it applied to the diameter, the radius, or—perhaps exceptionally—the circumference. In the same way in radiation protection one should expect to attribute a quantity of 10^{-2} J kg^{-1} to either dose equivalent or absorbed dose. All too frequently it tends to be attributed to the non-specific quantity 'dose'. It is to help avoid this ambiguity that different special names have been given to the J kg^{-1} when used in conjunction with absorbed dose (specific energy, kerma and absorbed dose index) on the one hand and with dose equivalent (and dose equivalent index and effective dose equivalent) on the other. The name of the unit then helps to specify the quantity involved.

A.3 Graphs and Tables

Graphs are frequently used to show the relationship between quantities. Normally it is numerical values of those quantities that are marked on the axes of the graphs. It follows from equation (A.1) that each axis of a graph should therefore be labelled by a quantity divided by a unit, e.g. Absorbed dose/Gy. This is the system used by the Royal Society and the ICRU and adopted in this book. When a logarithm is plotted it must be the logarithm of a number as it is not possible to have the logarithm of a physical quantity. The scale should therefore be labelled with the logarithm of a quantity divided by a unit, such as log(activity/Bq) or log(absorbed dose rate/Gy s^{-1}).

Similarly in tabulations of numerical data the headings of columns in the tables should be something that is a pure number, namely the quantity divided by the unit in which it is expressed.

A.4 Quantities and Units for General Use

Table A.1 Radiation quantities and units for general use.

Quantity		Unit symbols		
Name	Symbol	SI	SI restricted name†	Special‡
Particle number	N	1		
Radiant energy	R	J		
(Particle) fluence	Φ	m^{-2}		
Energy fluence	Ψ	$J\,m^{-2}$		
(Particle) fluence rate§	ϕ	$m^{-2}\,s^{-1}$		
Energy fluence rate	ψ	$W\,m^{-2}$		
Cross section	σ	m^2		b
Mass attenuation coefficient	μ/ρ	$m^2\,kg^{-1}$		
Mass energy transfer coefficient	μ_{tr}/ρ	$m^2\,kg^{-1}$		
Mass energy absorption coefficient	μ_{en}/ρ	$m^2\,kg^{-1}$		
Total mass stopping power	S/ρ	$J\,m^2\,kg^{-1}$		$eV\,m^2\,kg^{-1}$

Linear energy transfer	L_Δ	J m⁻¹		eV m⁻¹ ‖
Radiation chemical yield	$G(x)$	mol J⁻¹		
Mean energy expended per ion pair	W	J		eV
Energy imparted	ε	J		
Specific energy (imparted)	z	J kg⁻¹	Gy	rad
Absorbed dose	D	J kg⁻¹	Gy	rad
Absorbed dose rate§	$\dot{D}$	J kg⁻¹ s⁻¹	Gy s⁻¹	rad s⁻¹
Kerma	K	J kg⁻¹	Gy	rad
Kerma rate§	$\dot{K}$	J kg⁻¹ s⁻¹	Gy s⁻¹	rad s⁻¹
Exposure	X	C kg⁻¹		R
Exposure rate§	$\dot{X}$	C kg⁻¹ s⁻¹		R s⁻¹
Activity§	A	s⁻¹	Bq	Ci
Air kerma-rate constant§	Γ_δ	m² J kg⁻¹	m² Gy Bq⁻¹ s⁻¹	m² rad Ci⁻¹ s⁻¹

† The symbol for the special name of the SI unit restricted to specified quantities.

‡ One should not infer that the size of a unit given in this column is equal to the size of a unit on the same line in the other columns.

§ Day (d), hour (h), and minute (min) may be used instead of second (s).

‖ For practical purposes, keV μm⁻¹ may be used as a convenient multiple.

From ICRU (1980).

A.5 Quantities and Units for Radiation Protection

Table A.2 Quantities and Units for Use in Radiation Protection.

Quantity		Unit symbols		
Name	Symbol	SI	SI restricted name†	Special‡
Dose equivalent	H	$J\,kg^{-1}$	Sv	rem
Dose equivalent rate§	$\dot{H}$	$J\,kg^{-1}\,s^{-1}$	$Sv\,s^{-1}$	$rem\,s^{-1}$
Absorbed dose index	D_i	$J\,kg^{-1}$	Gy	rad
Absorbed dose index rate§	$\dot{D}_i$	$J\,kg^{-1}\,s^{-1}$	$Gy\,s^{-1}$	$rad\,s^{-1}$
Dose equivalent index	H_i	$J\,kg^{-1}$	Sv	rem
Dose equivalent index rate§	$\dot{H}_i$	$J\,kg^{-1}\,s^{-1}$	$Sv\,s^{-1}$	$rem\,s^{-1}$
Shallow dose equivalent index	$H_{i,s}$	$J\,kg^{-1}$	Sv	rem
Deep dose equivalent index	$H_{i,d}$	$J\,kg^{-1}$	Sv	rem
Ambient dose equivalent	$H^*(d)$	$J\,kg^{-1}$	Sv	rem
Directional dose equivalent	$H'(d)$	$J\,kg^{-1}$	Sv	rem
Individual dose equivalent, penetrating	$H_p(d)$	$J\,kg^{-1}$	Sv	rem
Individual dose equivalent, superficial	$H_s(d)$	$J\,kg^{-1}$	Sv	rem

† The symbol for the special name of the SI unit restricted to specified quantities.
‡ One should not infer that the size of a unit given in this column is equal to the size of a unit on the same line in the other columns.
§ Day (d), hour (h), and minute (min) may be used instead of second (s).
From ICRU (1980, 1985).

A.6 Relationships between Radiation Quantities

(The relationships are shown for monoenergetic particles. For heterogeneous radiation weighted mean interaction coefficients or energies are required.)

(1) Energy fluence, $\Psi = $ Fluence, $\Phi \times$ Energy, E
(2) Exposure, $X = \Psi(\mu_{en}/\rho)_{air} \, e/W_{air}$
(3) Kerma, $K_m = \Psi(\mu_{tr}/\rho)_m$
(4) Absorbed Dose, $D = \lim_{m \to 0} \bar{z}$
(5) $\qquad\qquad D_m = \Psi(\mu_{en}/\rho)_m$
(6) $\qquad\qquad D_m = [X(\mu_{en}/\rho)_m/(\mu_{en}/\rho)_{air}] \, W/e$
(7) $\qquad\qquad D_m = K_m (1 - g) = K_m(\mu_{en}/\rho)_m/(\mu_{tr}/\rho)_m$
(8) $\qquad\qquad D = \Phi \, S_{col}/\rho$
(9) Dose Equivalent, $H = D \, Q \, N$
(10) Effective Dose Equivalent, $H_E = \Sigma w_T H_T$.

Equations (2), (5), (6), (7) and (8) require charged particle equilibrium.

References

AAPM 1980 *Rep.* 7, *Protocol for neutron beam dosimetry* (New York: Am. Inst. Phys.)
—— 1983 *Med. Phys.* **10** 741
Allisy A 1979 *Procès-Verbaux des séance du Comité International des Poids et Mesures* **47** 46
Almond P 1967 *Phys. Med. Biol.* **12** 13
Almond P R, Behmard M and Mendez A 1978a *Med. Phys.* **5** 63
Almond P R, Mendez A and Behmard M 1978b in *National and International Standardization of Radiation Dosimetry* vol II (Vienna: IAEA) p271
Almond P R and Svensson H 1977 *Acta Radiol. Ther. Phys. Biol.* **16** 177
Amols H I, Dicello J F, Awschalom M, Coulson L, Johnsen S W and Theus R B 1977 *Med. Phys.* **4** 486
Archer B R and Wagner L K 1982 *Med. Phys.* **9** 844
Aston G H and Attix F H 1956 *Acta Radiol.* **46** 747
Attix F H 1968 *Health Phys.* **15** 49
—— 1979 *Health Phys.* **36** 347
Attix F H, De La Vergne L and Ritz V H 1958 *J. Res. NBS* **60** 235
Auxier J A, Snyder W S and Jones T D 1968 in *Radiation Dosimetry* vol I ed F H Attix and W C Roesch (New York: Academic Press) p275
Awschalom M, Rosenberg I and TenHaken R K 1983 *Med. Phys.* **10** 307
Bach R L and Caswell R S 1968 *Radiat. Res.* **35** 1
Baird L C 1981 *Med. Phys.* **8** 319
Bambynek W, Craseman B, Fink R W, Freund H-U, Mark H, Swift C D, Price R E and Rao P V 1972 *Rev. Mod. Phys.* **44** 716
Barnard G P, Aston G H, Marsh A R S and Redding K 1964 *Phys. Med. Biol.* **9** 333
Barnard G P, Axton E J and Marsh A R S 1959 *Phys. Med. Biol.* **3** 366
Becker K 1973 *Solid State Dosimetry* (Cleveland Ohio: CRC Press)
Bentley R E, Jones J C and Lillicrap S C 1967 *Phys. Med. Biol.* **12** 301
Berger M J 1968 *J. Nucl. Med. Suppl.* **1** 15
Berger M J and Seltzer S M 1983 *NBSIR* 82-2550-A (Washington, DC: NBS)
Berger M J, Seltzer S M, Domen S R and Lamperti P J 1975 in *Biomedical Dosimetry* STI/PUB/401 (Vienna: IAEA)
Bethe H A and Ashkin J 1953 in *Experimental Nuclear Physics* (New York: Wiley) vol I part II p166
Bewley D K, McCollough E C, Page B C and Sakata S 1974 *Phys. Med. Biol.* **19** 831
Birch R, Marshall M and Ardran G M 1979 *Catalogue of Spectral Data for Diagnostic X rays, Rep. No.* 30 (London: Hospital Physicists' Association)
Boag J W 1966 in *Radiation Dosimetry* vol II ed F H Attix and W C Roesch (New York: Academic Press) ch9

—— 1967 in *Solid State and Chemical Radiation Dosimetry in Medicine and Biology* (Vienna: IAEA) p349

Boag J W and Currant J 1980 *Br. J. Radiol.* **53** 471

Boag J W, Dolphin G W and Rotblat J 1958 *Rad. Res.* **9** 589

Bonnett D E and Parnell C J 1976 in *Basic Physical Data for Neutron Dosimetry* (Luxembourg: CEC) p43

—— 1982 *Br. J. Radiol.* **55** 48

Boutillon M, Henry W H and Lamperti P J 1969 *Metrologia* **5** 1

Boutillon M and Niatel M-T 1973 *Metrologia* **9** 139

Bragg W H 1912 *Studies in Radioactivity* (London: Macmillan)

Broerse J J 1980 *Ion chambers for neutron dosimetry* (Chur, Switzerland: Harwood Academic Publishers for CEC)

Broerse J J, Mijnheer B J and Williams J R 1981 *Br. J. Radiol.* **54** 882

Broszkiewicz R K and Bulhak Z 1970 *Phys. Med. Biol.* **15** 549

Brownell G L, Ellett W H and Reddy A R 1968 *J. Nucl. Med. Suppl.* **1** 27

Burch P R J 1955 *Radiat. Res.* **3** 361

Burger G, Morhart A, Nagarajan P S and Wittmann A 1984 private communication

Burhop E H S 1952 *The Auger Effect and other Radiationless Transitions* (London: Cambridge University Press)

Burke E A and Pettit R M 1960 *Radiat. Res.* **13** 271

Burlin T E 1961 *Phys. Med. Biol.* **6** 33

—— 1962 *Br. J. Radiol.* **35** 343

—— 1966 *Br. J. Radiol.* **39** 727

—— 1968 in *Radiation Dosimetry* vol. I ed F H Attix and W C Roesch (New York: Academic Press) p331

—— 1972 in *Topics in Radiation Dosimetry* ed F H Attix (New York: Academic Press) p144

Cairns J A, Holloway D F and Desborough C L 1969 *Nature* **223** 1262

Cameron J F and Ridley J D 1970 *IEEE Trans. Nucl. Sci.* **17** 363

Cameron J R, Suntharalingham N and Kenney G N 1968 *Thermoluminescent Dosimetry* (Madison, Wisconsin: University of Wisconsin Press)

Carlsson C A 1979 *Phys. Med. Biol.* **24** 1209

Caswell R S and Coyne J J 1972 *Radiat. Res.* **57** 448

Caswell R S, Coyne J J and Randolph M L 1980 *Radiat. Res.* **83** 217

Chan F K and Burlin T E 1970 *Health Phys.* **18** 325

Christen T 1913 *Messung und Dosierung der Röntgenstrahlen* (Hamburg: Graefe und Sillem)

—— 1914 *Arch. Roentgen Ray* **19** 210

Clark B C and Gros W 1968 *USAEC Rep.* N70 2740-5, p132

—— 1969 *Radiology* **93** 139

Cormack D V and Johns H E 1954 *Radiat. Res.* **1** 133

Cottens E, Janssens A, Eggermont G and Jacobs R 1981 in *Biomedical Dosimetry: Physical Aspects, Instrumentation, Calibration* (Vienna: IAEA) p189

Cross W G, Ing H and Freedman N 1983 *Phys. Med. Biol.* **28** 1251

Cunningham J R and Holt J G 1978 *Med. Phys.* **5** 64

Cunningham J R and Sontag M R 1980 *Med. Phys.* **7** 672

Curie P and Laborde A 1904 *C. R. Acad. Sci. Paris* **138** 1150

Davies J V and Law J 1963 *Phys. Med. Biol.* **8** 91

Dillman L T and Von der Lage F C 1975 *Radionuclide Decay Schemes and Nuclear Parameters for use in Radiation Dose Estimation, NM/MIRD Pamphlet No.* 10 (New York: Society of Nuclear Medicine)

Domen S R 1980 *Med. Phys.* **7** 157

—— 1982 *J. Res. NBS* **87** 211

—— 1983a *Int. J. App. Rad. Isot.* **34** 643

—— 1983b *Int. J. App. Rad. Isot.* **34** 927

Domen S R and Lamperti P J 1974 *J. Res. NBS* **78A** 595

Drexler G and Gossrau M 1968 *Spektren gefilterter Röntgenstrahlungen für Kalibrierzurecke: Ein Katalog, GSF Ber.* 545 (Neuherberg bei Munich: Institut Strahlenschutz)

Drexler G and Gossrau M 1968 *Spektren gefilterter Röntgenstrahlungen für Kalibrierzurecke: Ein Katalog, GSF Ber.* 545 (Neuherberg bei Munich: Institut Strahlenschutz)

Dudley R A 1966 in *Radiation Dosimetry* vol II ed F H Attix and W C Roesch (New York: Academic Press) ch15

Dutreix A and Wambersie A 1975 *Br. J. Radiol.* **48** 1034

Dyson N A 1973 *X Rays in Atomic and Nuclear Physics* (London: Longman) ch6

Eggermont G, Buysse J, Janssens A, Thielens G and Jacobs R 1978 in *National and International Standardization of Radiation Dosimetry* vol II (Vienna: IAEA) p317

Ellett W, Callahan A and Brownell G L 1964 *Br. J. Radiol.* **37** 45

—— 1965 *Br. J. Radiol.* **38** 541

Ellis S C 1977 in *Ionizing Radiation Metrology* ed E Casnati (Bologna: Editrice Compositori) p163

Emery E W 1966 in *Radiation Dosimetry* vol II ed F H Attix and W C Roesch (New York: Academic Press) p73

Epp E R and Weiss H 1966 *Phys. Med. Biol.* **11** 225

Epp E R, Weiss H, Ling C C, Djordjevic B and Kessaris N D 1975 in *Fast Processes in Radiation Chemistry and Biology* (Bristol: Inst. Phys. and New York: Wiley) p341

Euratom 1983 *Proc. 8th Symp. on Microdosimetry* (Luxembourg: CEC)

Eve A S 1906 *Phil. Mag.* **12** 189

Fano U 1954 *Radiat. Res.* **1** 237

Feist H 1982 *Phys. Med. Biol.* **27** 1435

Feist H, Harder D and Metzner R 1968 *Nucl. Instrum. Meth.* **58** 236

Fowler J F 1966 in *Radiation Dosimetry* vol II ed F H Attix and W C Roesch (New York: Academic Press) p291

Fricke H and Hart E J 1966 in *Radiation Dosimetry* vol II ed F H Attix and W C Roesch (New York: Academic Press) ch12

Fricke H and Morse S 1927 *Am. J. Roentgenol.* **18** 430

Genna S and Laughlin J S 1955 *Radiology* **65** 394

—— 1956 *Radiat. Res.* **5** 604

Gibb R and Massey J B 1980 *Br. J. Radiol.* **53** 1100

Gilfrich J V and Birks L S 1968 *Anal. Chem.* **40** 1077

Goodman L J 1978 *Proc. 3rd Symp. on Neutron Dosimetry in Biology and Medicine* ed G Burger and H Ebert (Luxembourg: CEC) p61

Goodman L J and Coyne J J 1980 *Radiat. Res.* **82** 13

Gossrau M and Drexler G 1971 in *Advances in Physical and Biological Radiation Detectors* (Vienna: IAEA) p205

Grant W H, Cundiff J H, Gagnon W F, Hanson W F and Shalek R J 1977 *Med. Phys.* **4** 68

Graves R G, Smathers J B, Almond P R, Grant W H and Otte V A 1979 *Med. Phys.* **6** 123

Gray L H 1929 *Proc. R. Soc.* A **122** 647

—— 1936 *Proc. R. Soc.* A **156** 578

—— 1944 *Proc. Camb. Phil. Soc.* **40** 72

Gray L H and Read J 1939 *Nature* **144** 439

Greene D 1962 *Phys. Med. Biol.* **7** 213

Greening J R 1950 *Proc. Phys. Soc.* **A63** 1227

—— 1951 *Proc. Phys. Soc.* **B64** 977

—— 1954 *Br. J. Radiol.* **27** 163

—— 1957 *Br. J. Radiol.* **30** 254

—— 1960 *Br. J. Radiol.* **33** 178

—— 1963 *Br. J. Radiol.* **36** 363

—— 1964 *Phys. Med. Biol.* **9** 143 (for correction see *Phys. Med. Biol.* **10** 566)

—— 1972 in *Topics in Radiation Dosimetry* ed F H Attix (New York: Academic Press) p261

—— 1983 in *Applications of Physics to Medicine and Biology* ed Alberi G, Bajzer Z and Baxa P (Singapore: World Scientific Publishing Co) p69

Greening J R and Randle K J 1968 *Phys. Med. Biol.* **13** 159

Greening J R, Randle K J and Redpath A T 1968a *Phys. Med. Biol.* **13** 359

—— 1968b *Phys. Med. Biol.* **13** 635

—— 1969 *Phys. Med. Biol.* **14** 55

Guiho J-P, Simoen J-P and Domen S R 1978 *Metrologia* **14** 63

Gunn S R 1964 *Nucl. Instrum. Meth.* **29** 1

—— 1970 *Nucl. Instrum. Meth.* **85** (1970 suppl.) 285

—— 1976 *Nucl. Instrum. Meth.* **135** 251

Hall E J and Oliver R 1961 *Br. J. Radiol.* **34** 397

Hannan W J, Porter D, Lawson R C and Railton R 1973 *Phys. Med. Biol.* **18** 808

Harder D 1968 *Biophysik* **5** 157

—— 1974 *Proc. 4th Symposium on Microdosimetry* (Brussels: CEC) p677

Haybittle J L 1960 *Br. J. Radiol.* **33** 52

Heintz P H, Johnsen S W and Peek N F 1977 *Med. Phys.* **4** 250

Henry W H 1979 *Phys. Med. Biol.* **24** 37

Herz R H 1969 *The Photographic Action of Ionising Radiations* (New York: Wiley-Interscience)

Hink W, Scheit A N and Ziegler A 1970 *Nucl. Instrum. Meth.* **84** 244

Hochanadel C J and Ghormley J A 1953 *J. Chem. Phys.* **21** 880

Holt J G, Fleischman R C, Perry D J and Buffa A 1979 *Med. Phys.* **6** 280

Horowitz Y S 1981 *Phys. Med. Biol.* **26** 765

HPA 1973 *Rep. Ser. No. 9, Diagnostic X-Ray Protection* (London: Hospital Physicists' Association)

—— 1983 *Phys. Med. Biol.* **28** 1097

Huang P-H, Kase K R and Bjärngard B E 1981 *Med. Phys.* **8** 368

—— 1982 *Med. Phys.* **9** 695

Hubbell J H 1969 *National Bureau of Standards Report* NSRDS-NBS 29

—— 1977 *Radiat. Res.* **70** 58

—— 1982 *Int. J. App. Rad. Isot.* **33** 1269

Hubbell J H, Veigele W J, Briggs E A, Brown R T, Cromer D T and Howerton R J 1975 *J. Phys. Chem. Ref. Data* **4** 471

Ing H and Cross W G 1975 *Phys. Med. Biol.* **20** 906

ICRP 1975 *Publ.* 23, *Report of the Task Group on Reference Man* (Oxford: Pergamon)

—— 1977 *Publ.* 26, *Recommendations of the Int. Comm. on Radiological Protection, Ann. ICRP* **1** No. 3

—— 1978 *Statement from the 1978 Stockholm Mtg of the Int. Commission on Radiological Protection, Ann. ICRP* **2** No. 1

—— 1983 *Publ.* 38, *Radionuclide Transformations* (Oxford: Pergamon)

ICRU 1938 *Br. J. Radiol.* **10** 438

—— 1951 *Br. J. Radiol.* **24** 54

—— 1954 *Br. J. Radiol.* **27** 243

—— 1957 *Rep.* 8 (Washington, DC: ICRU Publications)

—— 1961 *Rep.* 9 (Washington, DC: ICRU Publications)

—— 1962 *Rep.* 10a (Washington, DC: ICRU Publications)

—— 1968 *Rep.* 11, *Radiation Quantities and Units* (Washington, DC: ICRU Publications)

—— 1969a *Rep.* 13, *Neutron Fluence, Neutron Spectra and Kerma* (Washington, DC: ICRU Publications)

—— 1969b *Rep.* 14, *Radiation Dosimetry: X Rays and Gamma Rays with Maximum Photon Energies Between 0.6 and 50 MeV* (Washington, DC: ICRU Publications)

—— 1970a *Rep.* 16, *Linear Energy Transfer* (Washington, DC: ICRU Publications)

—— 1970b *Rep.* 17, *Radiation Dosimetry: X Rays Generated at Potentials of 5 to 150 kV* (Washington, DC: ICRU Publications)

—— 1971a *Rep.* 19, *Radiation Quantities and Units* (Washington, DC: ICRU Publications)

—— 1971b *Rep.* 20 *Radiation Protection Instrumentation and Its Application* (Washington, DC: ICRU Publications)

—— 1972 *Rep.* 21, *Radiation Dosimetry: Electrons with Initial Energies Between 1 and 50 MeV* (Washington, DC: ICRU Publications)

—— 1976 *Rep.* 25, *Conceptual Basis for the Determination of Dose Equivalent* (Washington, DC: ICRU Publications)

—— 1977 *Rep.* 26, *Neutron Dosimetry for Biology and Medicine* (Washington, DC: ICRU Publications)

—— 1979a *Rep.* 31, *Average Energy Required to Produce an Ion Pair* (Washington, DC: ICRU Publications)

—— 1979b *Rep.* 32, *Methods of Assessement of Absorbed Dose in Clinical Use of Radionuclides* (Washington, DC: ICRU Publications)

—— 1980 *Rep.* 33, *Radiation Quantities and Units* (Washington, DC: ICRU Publications)

—— 1982 *Rep.* 34, *The Dosimetry of Pulsed Radiation* (Washington, DC: ICRU Publications)

—— 1984a *Rep.* 35, *Radiation Dosimetry: Electron Beams with Energies Between 1 and 50 MeV* (Washington, DC: ICRU Publications)

—— 1984b *Rep.* 36, *Microdosimetry* (Washington, DC: ICRU Publications)

—— 1984c *Rep.* 37, *Stopping Powers for Electrons and Positrons* (Washington, DC: ICRU Publications)

—— 1985 *Rep.* 39, *Determination of Dose Equivalents for External Sources* (Washington, DC: ICRU Publications)

Jayaraman S, Lanzl L H and Agarwal S K 1983 *Med. Phys.* **10** 871
Jayaraman S, Sasane J B, Viswanathan P S and Ravichandran R 1979 *Med. Phys.* **6** 158
Johansson K-A, Mattsson L O, Lindberg L and Svensson H 1978 in *National and International Standardization of Radiation Dosimetry* vol II (Vienna: IAEA) p243
Johns H E and Laughlin J S 1956 in *Radiation Dosimetry* ed G J Hine and G L Brownell (New York: Academic Press) p73
Johnsen S W 1977 *Med. Phys.* **4** 255
Kaye G W C and Binks W 1937 *Proc. R. Soc.* A **161** 564
Kemp L A W 1977 in *Ionizing Radiation Metrology* ed E Casnati (Bologna: Editrice Compositori) p85
Kemp L A W, Marsh A R S and Baker M J 1971 *Nature* **230** 41
Kessaris N D 1970 *Radiat. Res.* **43** 281
King S D and Spiers F W 1985 *Br. J. Radiol.* in press
Knitter H H, Paulsen A, Liskien H and Islam M M 1973 *Atomkernenergie* **22** 84
Koch H W and Motz J W 1959 *Rev. Mod. Phys.* **31** 920
Kossel W, Mayer V and Wolf H C 1954 *Naturwissenschaften* **41** 209
Kramers H A 1923 *Phil. Mag.* **46** 836
Kretscho J, Harder D and Pohlit W 1962 *Nucl. Instrum. Meth.* **16** 29
Lamperti P J and Wyckoff H O 1965 *J. Res. NBS* **69C** 39
Laughlin J S 1969 in *Radiation Dosimetry* vol III ed F H Attix and E Tochilin (New York: Academic Press) p91
Laughlin J S and Genna S 1966 in *Radiation Dosimetry* vol II ed F H Attix and E Tochilin (New York: Academic Press) p389
Laurence G C 1937 *Can. J. Res.* **A15** 67
Law J and Redpath A T 1968 *Phys. Med. Biol.* **13** 371
Lederer C M, Hollander J and Perlman I 1967 *Table of Isotopes* (New York: Wiley)
Lederer C M and Shirley V S (eds) 1978 *Table of Isotopes* 7th edn (New York: Wiley)
Leonard B E and Boring J W 1973 *Radiat. Res.* **55** 1
Levy L B, Waggener R G, McDaris W D and Payne W H 1974 *Med. Phys.* **1** 62
Liesem H 1976 *Phys. Med. Biol.* **21** 360
Loevinger R 1969 in *Radiation Dosimetry* vol III ed F H Attix and E Tochilin (New York: Academic Press) ch18
Loevinger R and Berman M 1968 *Phys. Med. Biol.* **13** 205
Loevinger R, Holt J G and Hine G J 1956 in *Radiation Dosimetry* ed G J Hine and G L Brownell (New York: Academic Press) ch17
Loftus T P and Weaver J T 1974 *J. Res. NBS* **78A** 465
Lone M A, Alexander T K, Bigham C B, Ferguson A J, Fraser J A, McDonald A B and Schneider H R 1977 *Nucl. Instrum. Meth.* **143** 331
Lone M A, Ferguson A J and Robertson B C 1981 *Nucl. Instrum. Meth.* **189** 515
McDonald J C, Laughlin J S and Freeman R E 1976 *Med. Phys.* **3** 80
McKinlay A F 1981 *Thermoluminescence Dosimetry* (Bristol: Adam Hilger)
Madey R, Waterman F M and Baldwin A R 1977 *Med. Phys.* **4** 322
Mandour M A and Harder D 1977 in *Wissenschaftliche Tagung der Deutscher Gesellschaft für Medizinische Physik, Heidelberg* vol II ed W J Lorenz (Heidelberg: Hüthig) p291
Marinelli L D 1942 *Am. J. Roentgenol. and Rad. Therapy* **47** 210
Marinelli L D, Quimby E H and Hine G J 1948 *Am. J. Roentgenol. and Rad. Therapy* **59** 260
Mattsson L O, Johansson K-A and Svensson H 1981 *Acta Rad. Oncol.* **20** 385

Mayneord W V 1950 *Some Applications of Nuclear Physics to Medicine, Br. J. Radiol.* suppl. 2

Meredith W J 1967 *Radium Dosage: The Manchester System* (Edinburgh: Livingstone)

Meulders J P, Seleux P, Macq P C and Pirart C 1975 *Phys. Med. Biol.* **20** 235

Mijnheer B J, Haringa H, Nolthenius H J and Zijp W L 1981 *Phys. Med. Biol.* **26** 641

Mijnheer B J, Williams J R and Broerse J J 1981 in *Biomedical Dosimetry: Physical Aspects, Instrumentation, Calibration* (Vienna: IAEA) p329

Mika N and Reiss K H 1968 *Röntgenpraxis* **21** 164

—— 1969a *Strahlentherapie* **138** 760

—— 1969b *Tabellen zur Röntgendiagnostik* (Erlangen: Siemens Aktiengesellschaft)

Millar R H and Greening J R 1974 *J. Phys. B: Atom. Molec. Phys.* **7** 2332

Morris W T and Owen B 1975 *Phys. Med. Biol.* **20** 718

Mosse D, Cance M, Steinschaden K, Chartier M, Ostrowsky A and Simoen J P 1982 *Phys. Med. Biol.* **27** 583

Mountford P J, Thomas B J and Fremlin J H 1976 *Br. J. Radiol.* **49** 630

Nachtigall D and Burger G 1972 in *Topics in Radiation Dosimetry* ed F H Attix (New York: Academic Press) p385

NACP 1980 *Acta Rad. Oncol.* **19** 55

—— 1981 *Acta Rad. Oncol.* **20** 401

Nahum A E 1975 *PhD Thesis* Edinburgh University

—— 1978 *Phys. Med. Biol.* **23** 24

Nahum A E and Greening J R 1976 *Phys. Med. Biol.* **21** 862

—— 1978 *Phys. Med. Biol.* **23** 894

NCRP 1974 *Rep.* 41, *Specification of gamma-ray brachytherapy sources* (Washington, DC: NCRP)

—— 1978 *Rep.* 57, *Instrumentation and Monitoring Methods for Radiation Protection* (Washington, DC: NCRP)

Niatel M T 1967 *Phys. Med. Biol.* **12** 555

O'Dell A A, Sandifer C W, Knowlen R B and George W D 1968 *Nucl. Instrum. Meth.* **61** 340

Orton C G 1966 *Phys. Med. Biol.* **11** 377

Parker R P and Morley B J 1966 in *Solid State and Chemical Radiation Dosimetry in Medicine and Biology* (Vienna: IAEA) p167

Parnell C J 1972 *Br. J. Radiol.* **45** 452

Paterson R and Parker H M 1934 *Br. J. Radiol.* **7** 592

—— 1938 *Br. J. Radiol.* **11** 252, 313

Peaple L H J and Burt A K 1969 *Phys. Med. Biol.* **14** 73

Petree B and Ward G 1962 *NBS Tech. Note* 163

Pohlit W 1969 *Ann. New York Acad. Sci.* **161** 119

Pruitt J S and Domen S R 1962 *J. Res. NBS* **66A** 371

Radak B and Marković V 1970 in *Manual on Radiation Dosimetry* ed N W Holms and R J Berry (New York: Dekker) p45

Raeside D E 1976 *Phys. Med. Biol.* **21** 181

Ramm W J 1966 in *Radiation Dosimetry* vol II ed F H Attix and W C Roesch (New York: Academic Press) p123

Rao I S S and Naik S B 1980 *Med. Phys.* **7** 196

Rawlinson J A and Johns H E 1973 *Am. J. Roentgenol.* **118** 919

Reiss K H and Steinle B 1973 *Phys. Med. Biol.* **18** 746

Ritz V H 1961 *Radiat. Res.* **15** 460
Roesch W C 1958 *Radiat. Res.* **9** 399
—— 1968 in *Radiation Dosimetry* vol I ed F H Attix and W C Roesch (New York: Academic Press) ch5
Rossi H H and Failla G 1956 *Nucleonics* **14** 32
Roux A M 1976 *Metrologia* **12** 65
Saylor W L 1969 *Phys. Med. Biol.* **14** 87
Scharf K and Lee R M 1962 *Radiat. Res.* **16** 115
Schett A, Okamoto K, Lesca L, Fröhner F H, Liskien H and Paulsen A 1974 *Compilation Threshold Reactions. Neutron Cross Sections for Neutron Dosimetry and other Applications* (OECD Nuclear Energy Agency: Neutron Data Compilation Centre)
Scott P M and Greening J R 1963 *Phys. Med. Biol.* **8** 51
Shalek R J and Stovall M 1968 *Am. J. Roentgenol. Rad. Ther. and Nucl. Med.* **102** 662
—— 1969 in *Radiation Dosimetry* vol III ed F H Attix and E Tochilin (New York: Academic Press) ch31
Shiragai A 1978 *Phys. Med. Biol.* **23** 245
Shonka F R, Rose J E and Failla G 1958 in *2nd UN Conference on Peaceful Uses of Atomic Energy* vol 21 (New York: United Nations) p184
Siegbahn K 1965 *Alpha-, Beta- and Gamma-Ray Spectroscopy* vol I (Amsterdam: North-Holland)
Sievert R 1921 *Acta Radiol.* **1** 89
—— 1923 *Acta Radiol.* **3** 268
—— 1930 *Acta Radiol.* **11** 249
—— 1932 *Acta Radiol. Suppl.* 14
Silberstein L 1932 *J. Opt. Soc. Am.* **22** 265
Smathers J B, Otte V A, Smith A R, Almond P R, Attix F H, Spokas J J, Quam W M and Goodman L J 1977 *Med. Phys.* **4** 74
Snyder W S, Ford M R, Warner G G and Fisher H L 1969 *J. Nucl. Med. Suppl.* **3** 5
Snyder W S, Ford M R, Warner G G and Watson S B 1975 *Nuclear Medicine/Medical Internal Radiation Dose Committee,* Pamphlet **11** (New York: Society for Nuclear Medicine)
Soole B W 1971 *Phys. Med. Biol.* **16** 427
Spencer L W 1965 *Radiat. Res.* **25** 352
Spencer L W and Attix F H 1955 *Radiat. Res.* **3** 239
Stahel E 1929 *Strahlentherapie* **33** 296
Sternheimer R M and Peierls R F 1971 *Phys. Rev.* **3** 3681
Storm E and Israel H I 1970 *Nucl. Data Tables* **A7**
Stoval M and Shalek R J 1968 *Am. J. Roentgenol. Rad. Ther. and Nucl. Med.* **102** 677
Sunderaraman V, Prasad M A and Vora R B 1973 *Phys. Med. Biol.* **18** 208
Svensson H and Brahme A 1979 *Acta Rad. Oncology* **18** 326
Svensson H and Pettersson S 1967 *Ark. Fys.* **34** 377
Taylor L S and Singer G 1940 *Am. J. Roentgenol.* **44** 428
Thomson J J and Rutherford E 1896 *Phil. Mag.* (5th Ser.) **42** 392
Twidell J W 1970 *Phys. Med. Biol.* **15** 529
Ullmann J L, Peek N, Johnsen S W, Raventos A and Meintz P 1981 *Med. Phys.* **8** 396
Unsworth M H and Greening J R 1970 *Phys. Med. Biol.* **15** 631
Villard P 1908 *Arch. d'Elec. Med.* **16** 692
Walter F J 1970 *IEEE Trans. Nucl. Sci.* **17** 196

Waterman F M, Kuchnir F T, Skaggs L S, Kouzes R T and Moore W H 1979 *Med. Phys.* **6** 432

Werle H and Bluhm H 1972 *J. Nucl. Energy* **26** 165

White D R 1977 *Phys. Med. Biol.* **22** 889

Whittaker B 1970 in *Manual on Radiation Dosimetry* ed N W Holms and R J Berry (New York: Dekker) p363

Whyte G N 1954 *Nucleonics* **12** 18

—— 1959 *Principles of Radiation Dosimetry* (New York: Wiley)

Williams G, Swanson W P, Kragh P and Drexler G 1983 private communication

Williams J R, Ryall R E and Bonnett D E 1982 *Phys. Med. Biol.* **27** 81

Wyard S J 1955 *Nucleonics* **13** (No. 7) 44

Wyckoff H O 1960 *J. Res. NBS* **64C** 87

Wyckoff H O and Attix F H 1957 *NBS Handbook* 64 (Washington, DC: NBS)

Zimmer K G 1938 *Strahlentherapie* **63** 517, 528

Zoetelief J, Engels A C, Broerse J J and Mijnheer B J 1980 *Phys. Med. Biol.* **25**

Index